MICROTECHNOLOGY AND MEMS

MICROTECHNOLOGY AND MEMS

Series Editors: H. Fujita   D. Liepmann

The series Microtechnology and MEMS comprises text books, monographs, and state-of-the-art reports in the very active field of microsystems and microtechnology. Written by leading physicists and engineers, the books describe the basic science, device design, and applications. They will appeal to researchers, engineers, and advanced students.

M. Graf · D. Barrettino
H. P. Baltes · A. Hierlemann

CMOS Hotplate
Chemical Microsensors

With 70 Figures, 6 in Color and 12 Tables

 Springer

Dr. Markus Graf
ETH Zürich
Physical Electronics Laboratory
Wolfgang-Pauli-Straße 16
HPT-H8
8093 Zürich, Switzerland
e-mail: markus.graf@alumni.ethz.ch

Dr. Diego Barrettino
EPF Lausanne
INF 333
Station 14
1015 Lausanne, Switzerland
e-mail: diego.barrettino@epfl.ch

Prof. Dr. Henry P. Baltes
ETH Zürich
Physical Electronics Laboratory
Wolfgang-Pauli-Straße 16
HPT-H6
8093 Zürich, Switzerland
e-mail: baltes@phys.ethz.ch

Prof. Dr. Andreas Hierlemann
ETH Zürich
Physical Electronics Laboratory
Wolfgang-Pauli-Straße 16
HPT-H4.2
8093 Zürich, Switzerland
e-mail: hierlema@phys.ethz.ch

Series Editors:

Prof. Dr. Hiroyuki Fujita
University of Tokyo, Institute of Industrial Science
4-6-1 Komaba, Meguro-ku, Tokyo 153-8505, Japan

Prof. Dr. Dorian Liepmann
University of California, Department of Bioengineering
6117 Echteverry Hall, Berkeley, CA 94720-1740, USA

Library of Congress Control Number: 2007920910

ISSN 1615-8326

ISBN 978-3-540-69561-5 Springer Berlin Heidelberg New York

Springer is a part of Springer Science+Business Media

springer.com

© Springer-Verlag Berlin Heidelberg 2007

Typesetting and Production: LE-TEX Jelonek, Schmidt & Vöckler GbR, Leipzig
Cover design: eStudio Calamar Steinen

Printed on acid-free paper   SPIN 11316732   57/3100/YL – 5 4 3 2 1 0

Preface

The book "CMOS Hotplate Chemical Microsensors" provides a comprehensive treatment of the interdisciplinary field of CMOS technology-based chemical microsensor systems, and, in particular, of microhotplate-based systems. The book is, on the one hand, targeted at scientists and engineers that are interested in getting first insights in the field of microhotplates and related chemical sensing, since all necessary fundamental knowledge is included. On the other hand, it also addresses experts in the field since it provides detailed information on all important issues related to realizing microhotplates and, specifically, microhotplate-based chemical sensors in CMOS technology. A large number of microhotplate realizations and integrated-sensor-system implementations illustrate the current state of the art and, at the same time, give an impression of the future potential of chemical microsensors in CMOS technology. Since microsensors produce "microsignals", sensor miniaturization without sensor integration is, in many cases, prone to failure. This book will help to reveal the benefits of using integrated electronics and CMOS-technology for developing microhotplates and the corresponding chemical microsensor systems and, in particular, the advantages that result from realizing monolithically integrated sensor systems comprising transducers and associated circuitry on a single chip.

After a brief introduction, the fundamentals of miniaturized metal-oxide-based chemical sensors are laid out, which include the description of microhotplate structures as well as basic information on the gas detection mechanisms of the predominantly used metal-oxide material, tin oxide. The next section includes different methods of modeling the thermal properties of microhotplates and strategies to include those models into circuitry simulation programs. These more fundamental sections are followed by an extensive description of different microhotplate structures that have been realized in CMOS technology. Thereafter, a comprehensive overview of monolithically integrated CMOS microsystems comprising hotplates and the necessary driving and signal conditioning circuitry is given. In a next chapter, the development of monolithic sensor arrays and fully developed microsystems with on-chip sensor control and standard interfaces is presented. The book is concluded by a short summary, a review of the commercialization potential and possible applications, and an outlook to future developments.

As with all interdisciplinary efforts, teamwork plays a central role for being successful. Therefore the authors are particularly grateful to several highly motivated and excellent coworkers, colleagues and former students, who contributed to the work that is the topic of this book. Christoph Hagleitner, Kay-Uwe Kirstein, Sadik Hafizovic, Stefano Taschini, Urs Frey, Yue Li and Martin Zimmermann gave valuable input on the circuitry design side. The authors are grateful to Simon Müller, Reinhold Jurischka and Philipp Käser for their contributions to the microhotplate development. Jan Lichtenberg, Frauke Greve and Petra Kurzawski provided micromachining and microtechnological input, and Wan Ho Song helped with the microsensor packaging. The authors are also grateful to Prof. Oliver Brand, who always was a valuable source of information on microtechnology and microfabrication.

Moreover, the authors are indebted to European collaboration partners, Dr. Udo Weimar and Dr. Nicolae Bârsan and their collaborators at the University of Tübingen, and to AppliedSensor GmbH, Reutlingen, namely Dr. Stefan Raible, Dr. Jürgen Kappler, and Dr. Andreas Krauss, who provided the different chemically sensitive metal oxides.

Financial support for the CMOS microhotplate-based chemical-sensor projects came from the European Union (Projects CMOSSens, IST-1999-10579 and ADA, IST-2000-28452), the Swiss "Bundesamt für Bildung und Wissenschaft" (BBW), and the Swiss "Commission for Technology and Innovation" (CTI).

Contents

Abbreviations

AAF:	anti-aliasing filter
A/D:	analog-digital
AC:	alternating current
ADC:	analog-to-digital converter
AHDL:	analog hardware description language
ASIC:	application-specific integrated circuit
BF:	buffer
CMOS:	complementary metal oxide semiconductor
CVD:	chemical vapor deposition
D/A:	digital-analog
DAC:	digital-to-analog converter
DC:	directed current
DDA :	differential difference amplifier
DIL:	dual in line
DRIE:	deep reactive-ion etching
ECE:	electrochemical etch stop
FEM:	finite element modeling
FPGA:	field-programmable gate array
HDL:	hardware description language
HF:	high frequency
IC:	integrated circuit
I^2C:	inter integrated circuit, inter-IC
LF:	low frequency
LNA:	low-noise amplifier
LOD:	limit of detection
LPCVD:	low-pressure chemical vapor deposition
LPF:	low-pass filter
LSB:	least significant bit/byte
MEMS:	micro-electro-mechanical system
µHP :	microhotplate
MOS:	metal oxide semiconductor

MOSFET:	metal oxide semiconductor field effect transistor
MSB:	most significant bit/byte
MUX:	multiplexer
NMOS:	n-channel metal oxide semiconductor
OPAM:	operational amplifier
PC:	personal computer
PECVD:	plasma-enhanced chemical vapor deposition
PID:	proportional integral differential
PMOS:	p-channel metal oxide semiconductor
ppm:	parts per million
ppb:	parts per billion
PTAT:	proportional to absolute temperature
RGTO:	rheotaxial growth and thermal oxidation
RIE:	reactive-ion etching
SEM:	scanning electron microscopy
SCL:	serial clock
SDA:	serial data
SNR:	signal-to-noise ratio
SOI:	silicon on insulator
TO:	transistor outline
USB:	universal serial bus
UV:	ultraviolet
VHDL:	very-high-speed integrated-circuit hardware description language
VOC:	volatile organic compound

1

Introduction

In recent years, there has been increasing interest and development efforts in miniaturizing gas sensors and systems. Particularly strong efforts have been made to monitor environmentally relevant gases like carbon-monoxide (CO), methane (CH_4) and ozone (O_3). Commonly used chemically sensitive materials for these target gases are wide-bandgap semiconducting oxides such as tin oxide, tungsten oxide or indium oxide, which are operated at elevated temperatures of 200–400 °C [1–3]. At those high temperatures, these oxides show considerable resistance changes upon exposure to a multitude of inorganic gases and volatile organics. The most prominent example is tin oxide (SnO_2), which shows large electrical resistance changes upon exposure to the above-mentioned gases at operating temperatures between 250 °C–350 °C and has been engineered to provide sufficient long-term stability [4–6]. The miniaturization efforts in the field of metal-oxide-based gas sensors follow several major trends:

(a) the development of micromachined sensor platforms [7–9],
(b) the micro- and nanotechnological fabrication of the sensing materials [10, 11], and
(c) the design and co-integration of application-specific circuits with the transducer leading to smart sensor systems [8, 12–14].

During the last years, so-called "microhotplates" (µHP) have been developed in order to shrink the overall dimensions and to reduce the thermal mass of metal-oxide gas sensors [7, 9, 15]. Microhotplates consist of a thermally isolated stage with a heater structure, a temperature sensor and a set of contact electrodes for the sensitive layer. By using such microstructures, high operation temperatures can be reached at comparably low power consumption (< 100 mW). Moreover, small time constants on the order of 10 ms enable applying temperature modulation techniques with the aim to improve sensor selectivity and sensitivity.

The development of microhotplates is strongly coupled to novel nanotechnological fabrication strategies as well as microdeposition and microstructuring techniques for the respective metal oxides.

To date, most microhotplate-based chemical sensors have been realized as multi-chip solutions with separate transducer and electronics chips [16–19]. The co-integration of circuitry with chemical sensors requires sensor design and microfabrication processes that are compatible with the chip- and microelectronics technology to be used. The currently dominant and well-established technology for integrated circuits is CMOS (Complementary Metal Oxide Semiconductor) technology. Consequently, there is ongoing efforts to use CMOS technology for fabricating chemical sensors, and several CMOS-based monolithic systems have been successfully realized [14, 20]. CMOS-based systems do not only feature small size, but they also offer low power consumption, and the option of batch fabrication at industrial standards and low costs. The co-integration of transducers and circuits provides on-chip amplification and conditioning of sensors signals, enables on-chip analog-to-digital conversion, and allows for using on-chip standard interfaces, which alleviates the packaging problem (less pins and connections). Drawbacks of using CMOS technology include a limited selection of materials and a predefined fabrication process for the CMOS part. Sensor-specific or transducer-specific materials and fabrication steps have to be, in most cases, introduced during the post-processing after the CMOS fabrication. The integration of microhotplates using CMOS technology is particularly challenging, since the metal-oxide operating temperatures of 250 °C–350 °C are much higher than the temperature specifications of common integrated circuits (between –40 °C and 150 °C). Nevertheless, microhotplates relying on CMOS technology with subsequent additional micromachining have been presented by several groups [21–23]. The devices presented in this book additionally include examples of newly developed sensor systems featuring micromachined sensors, embedded smart features and nanotechnologically fabricated sensitive layers.

After this introduction, a general overview on microhotplate-based metal oxide sensor systems is given Chap. 2. Thermal modeling, which supports the device design and the fundamental understanding of the microhotplate thermal characteristics is detailed in Chap. 3. Three different prototypes of microhotplates in CMOS technology are presented as part of a design and transducer toolbox in Chap. 4. The first device is a circular-shape microhotplate fabricated with a minimum of post-processing steps. The design is optimized for the use with a drop-coated thick film nanocrystalline tin-oxide layer. The second device is designed with the aim to overcome the on-chip temperature limit of 400 °C, which is the consequence of the CMOS-metallization. The third microhotplate features a MOS-transistor heater that substitutes the conventional resistive heating element. All three microhoplates were, for the first time, integrated in monolithic sensor systems. In the subsequent Chap. 5 two monolithic systems are presented. The first monolithic system includes the circular microhotplate, a digital temperature controller, advanced read-out circuitry and a digital interface. The second monolithic sensor system features a high-temperature microhotplate and fully differential analog circuitry thus increasing the potential operation temperature range. Finally, several fully integrated sensor arrays are described in Chap. 6. The first array comprises the circular microhotplate and co-integrated analog circuitry such as a temperature controller and sensor read-out circuitry. The microhotplates were then further miniaturized in order to reduce the overall power consumption. The second arrray chip features these smaller microhotplates and a differential mixed-signal architecture, which enables individual temperature control and sensor read-out. The last chip comprises three transistor-heated microhotplates monolithically integrated with a fully digital sensor architecture and a digital interface. This chip shows the advantages and potential of using the CMOS-MEMS approach for metal-oxide-based microsensors. The book finally concludes with an outlook discussing future development work and potential applications of the presented sensor systems (Chap. 7).

2

Miniaturized Metal-Oxide Sensors

2.1 Overview of Microhotplates for Gas Sensing Applications

As already mentioned in the introduction, a so-called microhotplate (μHP) for metal-oxide-based gas sensing consists of a thermally isolated stage fabricated using microtechnological processes. The integrated heating element provides the typical operation temperature on the order of several $100\,^{\circ}$C, and the temperature sensor measures the microhotplate temperature. Two or more electrodes are used to perform resistance or impedance measurements of the sensing material. The microhotplate becomes a chemical sensor through the deposition of a sensitive layer. Several reviews on micromachined metal-oxide sensors are available [7, 9]. The scheme in Fig. 2.1 gives an overview on the design and development considerations for monolithic sensor systems.

Microhotplates, however, are not only used for metal-oxide-based gas sensor applications. In all cases, in which elevated temperatures are required, or thermal decoupling from the bulk substrate is necessary, microhotplate-like structures can be used with various materials and detector configurations [25]. Examples include polymer-based capacitive sensors [26], pellistors [27–29], GasFETs [30, 31], sensors based on changes in thermal conductivity [32], or devices that rely on metal films [33,34]. Only microhotplates for chemoresistive metal-oxide materials will be further detailed here. The relevant design considerations will be addressed.

The basic components of a microhotplate-based sensor system are shown in the lower part of Fig. 2.1. They include the microhotplate and dedicated sensor electronics. As shown in Fig. 2.1 the electronics part may feature a temperature controller, read-out and measurement circuitry for the different sensor elements, or a first data processing stage. Another interesting feature is an embedded sensor interface.

In a monolithic system, electronics and sensor part are integrated on a single chip. Separate circuitry and microsensor chips are characteristic for hybrid systems, as it is indicated by the dashed separation line in Fig. 2.1. The advantages and issues with hybrid and monolithic systems are extensively discussed in literature [14], so that only a few key points are mentioned here. The main disadvantage of monolithic systems is the limitation with respect to available materials and microtechnological process

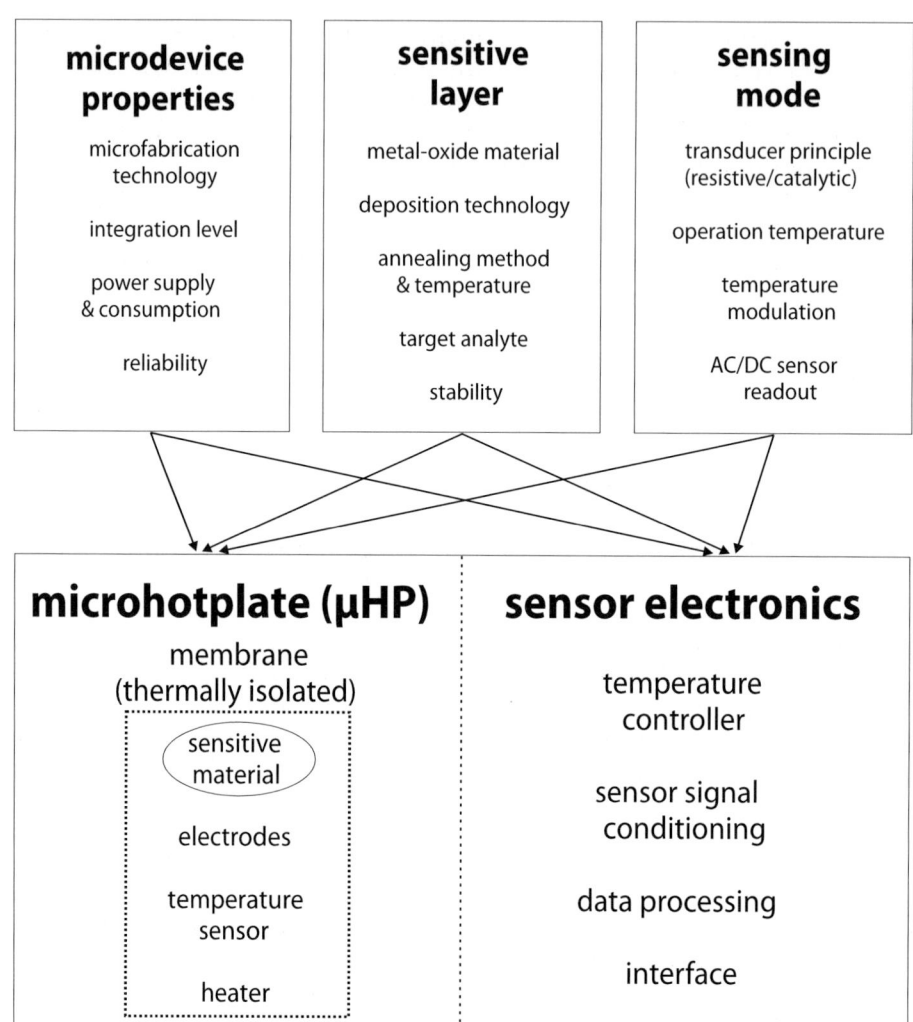

Fig. 2.1. Schematic diagram of design considerations for monolithic or hybrid metal-oxide-based sensor systems

steps, since the overall device fabrication has to be compatible with microelectronics fabrication processes as provided, e.g., by CMOS technology. Interdisciplinary expertise is required for system prototyping, and low-volume chip fabrication is expensive, so that the development of such systems is rather complex.

Monolithic sensor systems, however, offer advantages such as low power consumption or reduced packaging efforts. Integration of an on-chip serial interface reduces the number of bonding wires, especially in the case of an integrated multi-sensor array. Taking into account the mass-production facilities with established microelectronic and microtechnological processes, commercialization of monolithic systems appears to be very cost-effective. From the performance point of view, on-chip amplification and conditioning of small sensor signals is possible, and the miniaturized systems offer the advantage of reduced sensor size, and short recovery and response times, which render monolithic approaches to microhotplate-based sensing systems attractive. In most publications to date the microhotplates are realized, however, as discrete sensors that are read out by conventional measurement equipment or separate dedicated electronics.

In the upper part of Fig. 2.1, three main blocks are shown that represent all the issues related to the design and layout of microhotplate-based sensors and electronics. They include:

(a) microdevice properties
(b) the sensing layer material and its deposition technology
(c) the sensing mode and operational details.

First papers on microfabricated metal-oxide sensors were published by Demarne and Grisel [35]. The authors already used standard silicon technology such as thermally grown and chemical-vapor-deposited (CVD) silicon oxide as membrane material, as well as KOH-etching for membrane release. Over the last years, many more microhotplate types have been presented. New technologies were introduced, and the devices were optimized with respect to power consumption and response time. Besides closed membranes, spider-like structures were fabricated, the heated area of which is suspended by four arms, with the openings between the arms serving as etch-holes for frontside etching [21,36,37]. Removing some parts of the membrane increases the thermal resistance and decreases the total heat capacity. A recently presented device relies on SOI (Silicon-on-insulator) technology. A combination of front- and backside etching leads to bridge-like microhotplates suspended by only two arms [38]. A device with six legs including a novel SiC/HfB_2 thin film was recently presented [39].

Other devices feature a closed membrane made of dielectric layers such as silicon nitride or silicon oxide [40–44]. In some cases a thin Si-membrane is additionally included. The Si-membranes show good mechanical stability, but owing to the thermal conductivity of the highly doped silicon thin film serving as an etch-stop, the thermal resistance of the microhotplate is reduced [7,45]. Hence, a stress-optimized silicon oxinitride membrane has been used to replace the silicon layer in later designs [46]. A Si-island as heat spreader underneath the central heater area was shown to provide excellent temperature homogeneity [47]. The corresponding microhotplate was fabricated with a highly doped boron diffusion as etchstop in the island area. Such an island can be also created by a two-step backside etch in KOH thus avoiding additional diffusion steps [48]. Surface micromachining through sacrificial etching of either a polysilicon [49] or a porous silicon layer [29,50] has been used as an alternative membrane release process.

Many devices have been denoted to be "CMOS-compatible", this term, however, not being clearly defined. In most cases CMOS-compatible means, that CMOS materials have been used, or the design can be used within a modified CMOS-process. As modifications in industrial CMOS-processes are difficult to implement, two main approaches have been pursued so far. One approach relies on an open process window for the sensor, so that the post-process formation of the microhotplate is as much as possible decoupled from the CMOS-process. CMOS-compatible in this case means, that the post-processing steps do not require temperature steps at temperatures higher than 400 °C, since the CMOS metallization might degrade, and the device parameters may be altered. Therefore, LPCVD (Low Pressure Chemical Vapor Deposition) oxides and nitrides with typical deposition temperatures of 700–800 °C are not applicable [51]. The deposition and annealing temperatures of the sensitive layer are also limited. The sensor development is largely independent of the complex CMOS-process. Several groups pursued such an approach, but neither full process integration nor monolithic devices have been presented so far [23, 38].

The other method includes the pre-definition of the microhotplate in the CMOS-process. Additional post-processing steps are then used for the microhoplate formation. Examples are the microhotplates [21], which were developed at the National Institute for Standard and Technology (NIST, Gaithersburgh, MD, USA) and the microhotplates that will be discussed in this book. Another example is a microhotplate relying on SOI (Silicon On Insulator)-CMOS technology with a transistor instead of a resistor as heating element [22, 52]. All these devices have the potential for monolithic integration with circuitry, as will be discussed in one of the next paragraphs.

The second block in Fig. 2.1 is the sensitive layer. The discussion here is limited to sensitive layers on micromachined substrates, which can be categorized according to the deposition method. A more detailed discussion of various methods can be found in [53]. One set of methods relates to thin-film technology and includes the direct deposition by means of conventional microtechnological processes such as chemical-vapor deposition (CVD), sputtering or thermal evaporation. A general difficulty is the precise patterning of these materials at micrometer resolution. Lift-off techniques are applicable for a variety of metal oxides, but impose limits on the deposition temperature [23, 49]. For in-situ CVD-processes, the microhotplate is heated, which leads to the decomposition of the gaseous precursor material and a locally defined deposition of a thin-film layer in the heated area [54]. The hotplate temperature heavily influences the film morphology in CVD-processes, which also holds true for the case of sputtering. Masklessly sputtered films have not to be patterned according to the authors of [37], since the resistance of the non-heated material is very high, and, therefore, gas-sensitive electrical conduction effects will only occur in the heated parts, i.e., on the microhotplate. Thermolithographic patterning of sol-gel materials constitutes an alternative, which was reported on recently [55]. Another possibility includes rheotaxial growth and thermal oxidation of tin metal layers (RGTO): A tin layer is deposited and subsequently oxidized. The layer is patterned either by lift-off or by using a shadow mask [56, 57]. The result is a locally defined nanocrystalline thin-film layer.

The other category of deposition methods starts with the production of a metal-oxide powder. This powder is mixed with water and/or organic solvents to produce a paste, which is deposited onto the substrates. The deposition can either be done by spin-coating in combination with etching or by thermolithograhic patterning through membrane heating [58, 59]. There are also methods relying on pulverization deposition [60]. All these deposition technologies lead to nanocrystalline metal-oxide layers in the micrometer and submicrometer layer-thickness range. Even thicker layers (tens of microns) are fabricated by drop coating [30,61], screen printing [62,63] or fluidic paste deposition [64], for which post-structuring is not necessary.

The gas-sensing properties of the device, i.e., the metal-oxide conduction phenomena and mechanisms depend on the characteristic dimensions of the crystallites, which are typically on the order of 10 nm [65,66]. Thick-film sensing layers consist of highly porous agglomerates of crystallites that feature a large surface area. Co-integrated electronics on the device impose a temperature limit for film sintering and stabilization (the CMOS limit is, e.g., 400 °C). Such limits are not imposed on the pre-processing of the powders, but apply afterwards to the annealing of the paste on the membrane.

Layer doping with catalytic metals can be done either during the powder preparation or later in the deposition and annealing process. Doping is an important procedure in tuning the gas-sensor characteristics (selectivity pattern of the sensitive layers) [67,68].

The third block in Fig. 2.1 shows the various possible sensing modes. The basic operation mode of a micromachined metal-oxide sensor is the measurement of the resistance or impedance [69] of the sensitive layer at constant temperature. A well-known problem of metal-oxide-based sensors is their lack of selectivity. Additional information on the interaction of analyte and sensitive layer may lead to better gas discrimination. Micromachined sensors exhibit a low thermal time constant, which can be used to advantage by applying temperature-modulation techniques. The gas/oxide interaction characteristics and dynamics are observable in the measured sensor resistance. Various temperature modulation methods have been explored. The first method relies on a train of rectangular temperature pulses at variable temperature step heights [70–72]. This method was further developed to find optimized modulation curves [73]. Sinusoidal temperature modulation also has been applied, and the data were evaluated by Fourier transformation [75]. Another idea included the simultaneous measurement of the resistive and calorimetric microhotplate response by additionally monitoring the change in the heater resistance upon gas exposure [74–76].

Most microhotplate-based chemical sensors have been realized as multi-chip solutions with separate transducer and electronics chips. One example includes a gas sensor based on a thin metal film [16]. Another example is a hybrid sensor system comprising a tin-oxide-coated microhotplate, an alcohol sensor, a humidity sensor and a corresponding ASIC chip (Application Specific Integrated Circuit) [17]. More recent developments include an interface-circuit chip for metal oxide gas sensors and the conccept for an on-chip driving circuitry architecture of a gas sensor array [18,19].

The first monolithic devices have been presented at the same time by a group at NIST and a group at the Physical Electronics Laboratory (PEL) of ETH Zurich [77–81]. The NIST chip hosts an array of microhotplates integrated with transistor switches and a readout amplifier for the sensitive layer. The device presented by PEL includes an analog temperature controller and a logarithmic converter for reading out the sensor values. This was the first monolithic realization of an embedded system architecture with integrated microhotplate.

2.2 Microhotplates in Industrial CMOS Technology

A cross-sectional schematic of a monolithic gas sensor system featuring a microhot-plate is shown in Fig. 2.2. Its fabrication relies on an industrial CMOS-process with subsequent micromachining steps. Diverse thin-film layers, which can be used for electrical insulation and passivation, are available in the CMOS-process. They are de-noted "dielectric layers" and include several silicon-oxide layers such as the thermal field oxide, the contact oxide and the intermetal oxide as well as a silicon-nitride layer that serves as passivation. All these materials exhibit a characteristically low thermal conductivity, so that a membrane, which consists of only the dielectric layers, pro-vides excellent thermal insulation between the bulk-silicon chip and a heated area. The heated area features a resistive heater, a temperature sensor, and the electrodes that contact the deposited sensitive metal oxide. An additional temperature sensor is integrated close to the circuitry on the bulk chip to monitor the overall chip tempera-ture. The membrane is released by etching away the silicon underneath the dielectric layers. Depending on the micromachining procedure, it is possible to leave a silicon island underneath the heated area. Such an island can serve as a heat spreader and also mechanically stabilizes the membrane. The fabrication process will be explained in more detail in Chap 4.

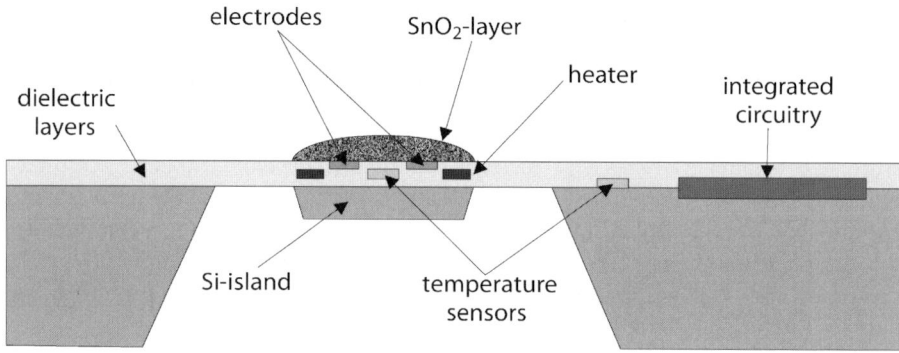

Fig. 2.2. Cross-sectional schematic of a monolithic sensor system in CMOS technology

2.3 Basic Sensing Mechanisms in Tin-Oxide Materials

2.3.1 Nanocrystalline Tin-Oxide Thick-Film Layers

The sensitive layers that have been used throughout this book consist of nanocrys-talline tin-oxide thick films. The resistance change is the result of a multitude of reac-tions taking place at the surface and in the bulk. This resistance change depends also on the morphology of the sensitive layer and the contact-electrode geometry. Due to the complex chemistry generating the sensor signal, a direct correlation with analyte concentrations is difficult. Sensor results have to be compared to results of other measurement methods, to reveal the reaction mechanisms and interactions that occur between sensing layers and analytes [65, 66]. Investigations on the fundamentals of the sensor processes are still in progress [82–84], the results of which will lead to further improvements in sensor technology.

The model analytes, which were used to show the sensor performance of the microsystems include carbon monoxide, CO, and methane, CH_4. The sensor microsystems were designed for practical applications, such as environmental monitoring, industrial safety applications or household surveillance, which implies that oxygen and water vapors are present under normal operating conditions. In the following, a brief overview of the relevant gas sensor mechanisms focused on nanocrystalline tin-oxide thick-film layers will be given.

Carbon-Monoxide Sensing

Figure 2.3 shows a schematic view of the nanocrystalline sensor material. It consists of single-crystalline tin-oxide grains with a typical size of 10 nm and a narrow size distribution [68]. The grains are in loose contact. The lower graph in Fig. 2.3 schematically represents the conduction band of the layer.

Typical operating temperatures are 200 °C to 400 °C. If oxygen is present, it is ionosorbed on the surface as O^- or O^{2-}. A layer of negative charges builds up at the grain surface and leads to a depletion of the conduction band at the surface.

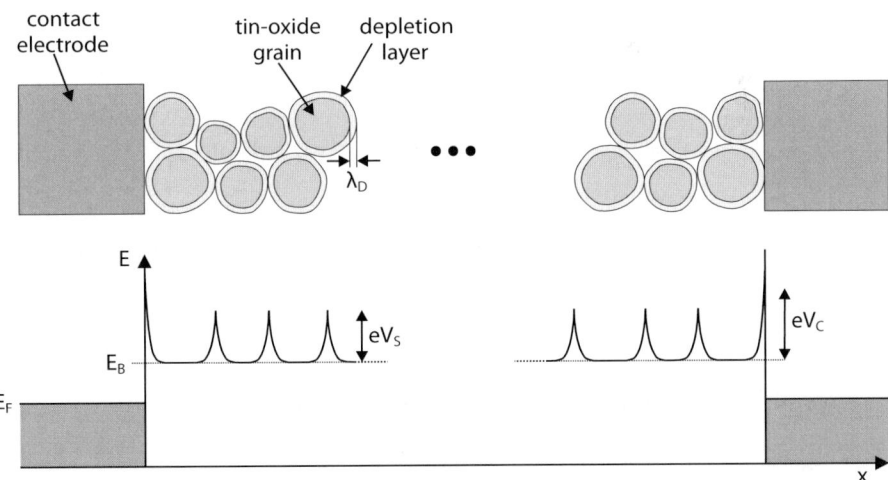

Fig. 2.3. Schematic view of a porous nanocrystalline sensing layer with a one-dimensional representation of the energetic conduction band. A inter-grain band bending, eV_S, occurs as a consequence of surface phenomena, and a band bending, eV_C, occurs at the grain-electrode contact. E_B denotes the minimum conduction band energy in the bulk tin oxide, and E_F is the Fermi-energy in the electrode metal

The Debye length, λ_D, is characteristic for the depletion depth. In Fig. 2.3, the grain size is considerably larger than the Debye length, which is represented by the light halo along the grain boundary. A typical value for λ_D is 20 nm at a temperature of 600 °K [65]. Under these conditions the nano-size grains (approx. diameter 10 nm) are fully depleted.

For conduction through the sensitive layer, the electrons have to pass the intergranular contacts and associated potential barriers. The height of these barriers depends on the composition of the ambient gas atmosphere.

Carbon monoxide molecules react with the adsorbed oxygen species at the surface of the grains and form carbon dioxide, CO_2, according to:

$$CO + O_{ad}^- \longrightarrow CO_2 + e^- \tag{2.1}$$

CO acts as a reducing gas and lowers the surface barriers between the grains, so that the resistance of the sensitive layer drops.

Experimental results and theoretical considerations suggest a power law for the dependence of the sensor resistance, R_S, on the CO partial pressure, p_{CO} [65]:

$$R_S \sim p_{CO}^{-n} \tag{2.2}$$

where n depends on materials properties such as surface morphology and state and on the electrode configuration. The negative exponent indicates that the sensor resistance decreases with increasing CO concentration.

Interaction With Water Vapor

The interaction of tin oxide and water also leads to a decrease of the sensor resistance, for which several reaction mechanisms were proposed. The first mechanism involves tin and oxygen lattice atoms reacting with water [1]:

$$H_2O + Sn_{lat} + O_{lat} \longleftrightarrow (HO - Sn_{lat}) + (O_{lat}H)^+ + e^- \tag{2.3}$$

The rooted OH group including the lattice oxygen serves as electron donor, whereas the hydroxyl group bound to the lattice tin forms a dipole.

A second possibility of electron donation is a reaction, in which a double-charged oxygen vacancy, V_O^{2+}, is created [1]:

$$H_2O + 2Sn_{lat} + O_{lat} \longleftrightarrow 2(HO - Sn_{lat}) + V_O^{2+} + 2e^- \tag{2.4}$$

Additional explanations consider the influence of adsorbed water molecules on the oxygen chemisorption, which affects the electronic barrier heights between the grains [85].

Influence of Humidity on the CO-Detection

The reaction mechanisms for water vapors do not only influence the sensor resistance but also the detection of CO. A promotion of CO-induced reactions is observed.

In both interaction mechanisms with water molecules as proposed by Eq. (2.3) and Eq. (2.4), hydroxyl groups are formed on the surface. These hydroxyl groups are additional reaction partners for CO. Starting from Eq. (2.4) and considering experimental findings, the reaction can be expressed as [86, 87]:

$$CO + (HO - Sn_{lat}) + O_{lat} \longleftrightarrow CO_2 + Sn_{lat} + (O_{lat}H)^{+} + e^{-} \qquad (2.5)$$

The oxygen vacancies created according to Eq. (2.4) act as additional adsorption sites for oxygen, which increases the concentration of adsorbed oxygen and results in a larger resistance change.

The following experimental observations qualitatively summarize the influence of humidity on the CO sensing process [66].

If humidity increases,

- R_S^{air} in air decreases
- R_S^{CO} for a defined CO concentration decreases
- the sensor signal R_S^{air}/R_S^{CO} upon CO exposure increases.

The sensor resistance can still be approximated by a power law as given by Eq. (2.2) with a specific exponent n.

Methane Sensing

Another analyte of interest is methane (CH_4). Its detection is highly relevant to safety applications such as natural-gas leakage detection or environmental monitoring. Methane is a known interferant to CO, because the tin-oxide resistance also decreases in the presence of methane [88]. The operating temperature for achieving large sensor signals upon the presence of methane is generally higher than for CO. The methane-induced change is attributed to reactions with lattice oxygen and ionosorbed oxygen. Both reaction paths lead from an adsorbed methyl group to a C-H group that is rooted on the tin-oxide surface [89].

In the presence of humidity, the sensor signal for a given methane concentration decreases. This can be attributed to competing adsorption of water and methane molecules at the same sites, which holds particularly true for lattice oxygen sites [90].

2.3.2 Catalyst Doping

It is well established, that the poor selectivity of tin-oxide sensors can partly be overcome by adding catalysts to the sensitive layer. Most common additives are noble metals like gold (Au), platinum (Pt) or palladium (Pd). They can be mixed with the tin oxide during paste formation before deposition. The influence of dopants on the gas sensor response is still subject to debates. The two most established mechanisms are the spill-over and the Fermi-level mechanism [82].

In the spill-over or catalytic model, the noble-metal clusters on the surface act as catalytic reaction sites (Fig. 2.4a). Reacting species such as oxygen can be dissociated more easily at these sites. When they move from the metal cluster to the grain

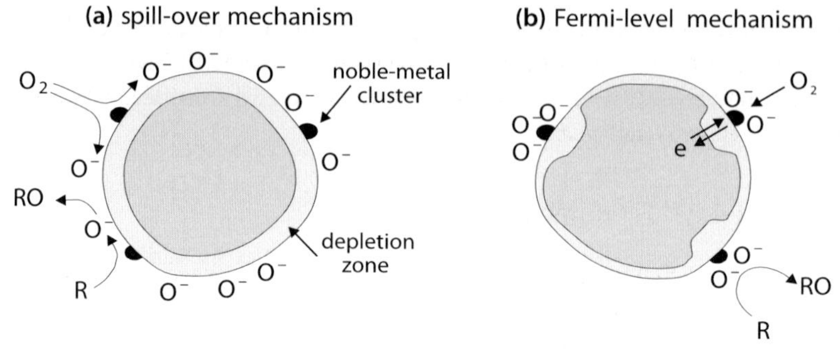

Fig. 2.4. Effects of noble-metal catalyst doping on the tin oxide: (**a**) Spill-over mechanism, (**b**) Fermi-level mechanism

surface (spill-over) additional adsorbed reactants are provided. Thus, the presence of the catalyst leads to a higher coverage of reactants on the grain surface. Moreover, the presence of the catalyst promotes and accelerates the reaction of the analyte with the adsorbed reaction partners, and an altered sensor response is observed. Higher selectivity is achieved, if the catalytic activation is specific for certain gases.

In the so-called "Fermi-level mechanism" it is assumed, that chemisorbed oxygen on the noble-metal cluster traps electrons that come from the semiconductor grain (Fig. 2.4b). This electron trapping implies a change in the depletion layer and in the band bending at the grain surface, thereby changing the conductivity of the nanocrystalline material. The oxidation state and the Fermi-level of the metal cluster depend on the composition of the ambient gas atmosphere. Hence, the influence of the gas atmosphere on the electronic properties of the metal cluster will alter the conduction in the sensitive layer owing to the above-mentioned electron trapping effect. The catalyst doping consequently can increase the selectivity of tin oxide towards certain analytes. Palladium is considered a typical representative of the Fermi-level mechanism.

The doping-induced effects on the gas sensing performance of nanocrystalline tin oxide include, first and foremost, a conductivity decrease in clean air by one to three orders of magnitudes, and, secondly, a shift of the optimal sensor working temperature from higher to lower temperatures.

Combustible gases will have a much higher reaction rate on Pd-doped layers due to catalytic effects. Two mechanism may help to explain the experimental findings: First, increased reaction rates lead to higher electron density in the surface area, thus generally increasing the sensor response. Second, the catalytic activity can increase the reaction rate for certain analytes in such a way that all available analyte molecules are consumed. Analytes with high reaction rates are completely consumed, and the gas sensing response is dominated by the species with lower reaction rates.

The model sensitive layer, which will be used for gas sensor performance tests throughout this book, was SnO_2 that has been doped with 0.2 wt % Pd. The minute Pd-content leads to a better sensitivity to carbon monoxide. The larger response is a consequence of the increased reaction rate. For the sensor arrays in Chap. 6, two additional materials have been prepared. Pure tin oxide shows a good sensor response to NO_2 and a lower sensitivity to CO. Pd-doping concentrations of 3% result in a better sensitivity to methane. Highly Pd-doped layers are operated at higher temperatures so that the catalytic additive helps to combust the carbon monoxide, which results in reduced cross-sensitivity to CO.

2.3.3 Sensitive-Layer Fabrication

The sensitive layers that have been used throughout this book were produced and deposited by AppliedSensor (AS, Reutlingen, Germany). Although metal-oxide-based microhotplate-sensors are already commercially available, a brief description of the paste production is given for the sake of completeness. The process is detailed in [82].

The fabrication starts with tin tetrachloride, $SnCl_4$, in an aqueous solution. Upon presence of NH_3, the SnO_2 precipitates as nanosize particles. The suspension is centrifuged and dried in an oven at 80 °C, so that hydrated SnO_2 remains, which is the precursor material for the calcination. The calcination takes places in an oven at elevated temperatures between 250 °C and 1000 °C. During heating the water is removed, and the crystallites grow in a controlled way. The product is a SnO_2 powder, that is ground in order to separate agglomerates and homogenize the powder.

The powder is then suspended in an aqueous solution of the doping metal chloride, e.g., $PdCl_2$, so that the metal chloride will attach to the tin-oxide crystallites. Annealing during 1 h at 450 °C releases the chloride, and the Pd-clusters remain at the grain surface. The powder is finally mixed with an organic solvent, and a homogeneous paste is produced. This paste is the starting material for the sensitive-layer deposition on the micromachined substrates.

3

Thermal Modelling of CMOS Microhotplates

3.1 Modelling Approach

The main goals of improving microhotplate designs include (a) reducing its power consumption and (b) increasing the hotplate temperature homogeneity, i.e., an optimization towards a minimum temperature gradient in the sensitive area. To reach these goals, novel device and hotplate designs usually undergo an extensive simulation process such as thermal modelling using finite-element simulations. Modelling the transducer and hotplate behavior is an important step towards establishing a compact sensor model for monolithic system realizations [91]. The parameters of interest include the hotplate thermal resistance and the thermal time constant, a prediction of which facilitates and accelerates the design of monolithic systems for a given microhotplate design.

Many of the hotplate devices presented so far rely on corresponding thermal simulations that are based on model assumptions and finite element methods (FEM) [47, 92–97]. Analytical models also have been developed [7,9,98,99]; another publication describes RC-network analysis and dimension reduction [100]. A reduction of the complexity and order of the model has been successfully realized, and the different relevant approaches have been summarized in recent articles [101, 102]

The general model assumptions and FEM implementations depend on the geometrical dimensions and the hotplate layouts. Most of the approaches are based on linear approximations, i.e., the temperature coefficients of the heat conductivity are not included. The temperature coefficients, however, are on the order of $10^{-3}/°C$ [103, 104], and will, depending on the geometry, noticeably influence the temperature distribution in the typical operating temperature range of 250–350 °C.

Another issue is how to include heat conduction and dissipation through ambient air. A heat transfer coefficient, h, is commonly used in 2-dimensional simulations, which is difficult to determine, since it is strongly depending on the package of the microhotplate sensor. It is, therefore, in most cases introduced as a parameter that is varied to fit the experimental data [45,94].

The specific heat of the ambient air is generally temperature-dependent, which entails a temperature-dependent h. The heat conductivities of the transducer thin film materials are, in most cases, not well known, and the thin-film properties can considerably differ from that of the bulk material. In case that the heat conductivity is not known, it can be implemented as a fitting parameter, although this might affect the validation of the model assumptions.

The focus of this chapter is on improving the thermal modelling of a microhotplate in CMOS technology. Thermal simulations of such microhotplates have two main purposes: (a) gaining information about the temperature distribution over a given microhotplate structure and facilitating its optimization procedure, and (b) providing input parameters for the overall microsystem simulation. The temperature dependence of the heat conductivities has to be included, which results in a nonlinear modelling problem. Another important guideline was to avoid the introduction of fit parameters.

A key issue in monolithic-system design is the simulation of the microhotplate coupled to the circuitry to ensure full chip functionality. This requires an adequate description of the microhotplate and an implementation in a language that is applicable to circuitry simulations.

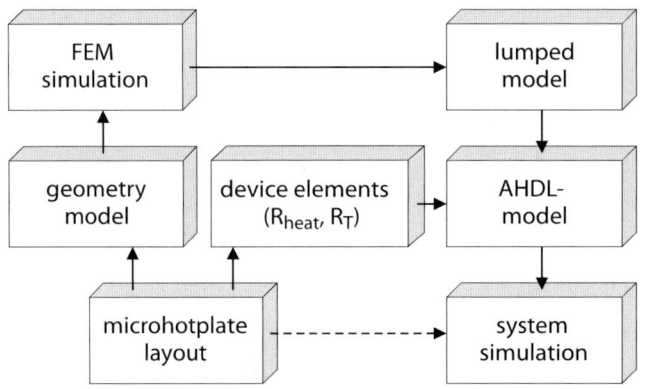

Fig. 3.1. Schematic of the modelling steps needed to come from a microhotplate layout to a model description for system-level simulations

The suggested procedure to arrive at this goal is presented in Fig. 3.1. It starts with the transfer of a certain microhotplate layout into a geometry model for a complex FEM simulation. This step is shown in Fig. 3.2 and will be explained in more detail in one of the next sections. A complex 3-d FEM simulation is then performed. The results of this simulation are used to produce a lumped-element model. This model is translated into a hardware description language (HDL). Using the resistances of the device elements such as the heater resistance, R_{heat}, and the resistance of the temperature sensor, R_T, co-simulations with the circuitry can be performed.

The aim in converting the microhotplate into a geometry model for the FEM simulation was to find a model that is as simple as possible but includes all relevant processes. The model assumptions to be explained in detail in the following section and the steps to arrive at the model are represented in Fig. 3.2. The feature on the bottom left-hand side represents the microhotplate schematic. The microhotplate exhibits a symmetric design so that a simulation of one quarter is adequate. Geomet-

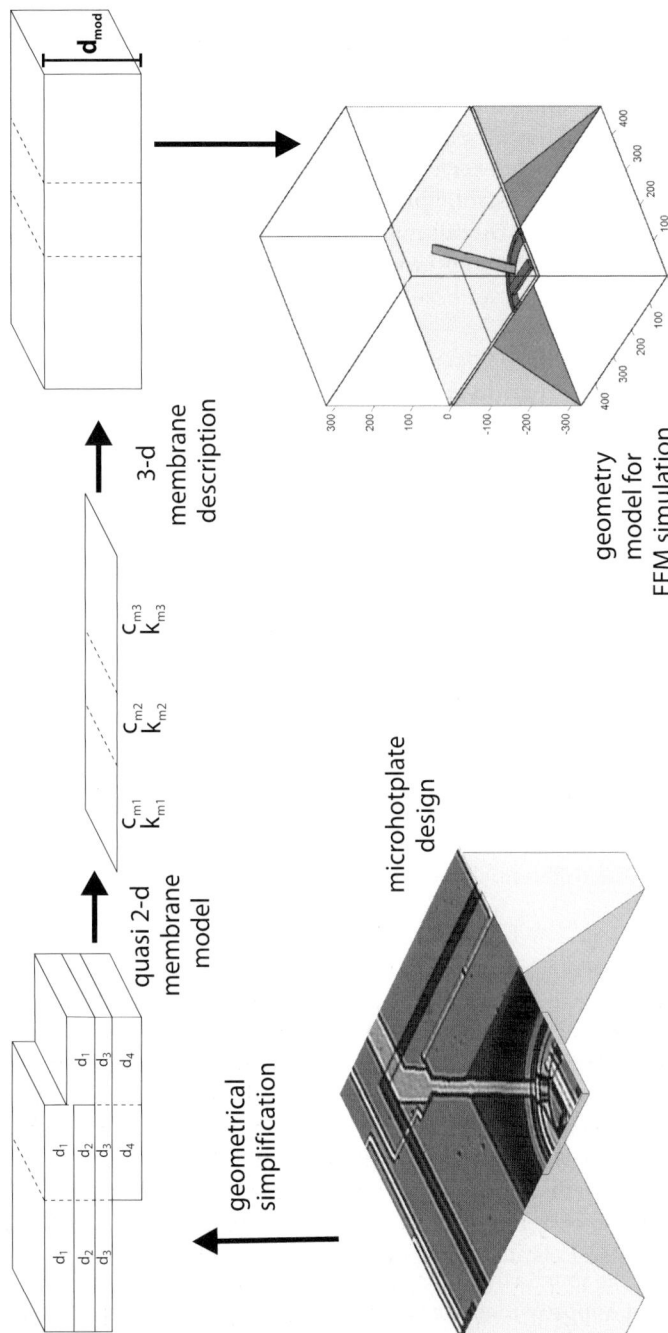

Fig. 3.2. Schematic showing the translation of a microhotplate layout into a geometry model description for FEM simulations

rical simplifications are introduced into the representation of the membrane layout and structure, for which a quasi 2-d membrane model is developed in Sect. 3.2. Afterwards, the membrane model is transferred back to a simplified 3-d description of the membrane, the different sections of which exhibit homogeneous heat conductivity and constant thickness. Finally, the membrane structure is combined with the supporting silicon frame and the surrounding air. The result is a geometry model for the FEM simulation as discussed in Sect. 3.3. The input parameters for a lumped-model description (Sect. 3.4) of the microhotplate are calculated from the simulation results. An AHDL (Analog Hardware Description Language) model, which then will be used for a circuitry simulation of the complete sensor system, is derived from the lumped-model equations in Sect. 3.5.

3.2 Microscopic Description and Model Assumptions

The governing equation for heat conduction with temperature-dependent heat conductivities can be written as:

$$j(\underline{x},t) = -\kappa(\underline{x},T(\underline{x},t)) \cdot \nabla T(\underline{x},t) \tag{3.1}$$

where $j(\underline{x},t)$ denotes the heat flux, and $\kappa(\underline{x},T(\underline{x},t))$ is the heat conductivity that locally depends on the temperature. $T(\underline{x},t)$ denotes the temperature field as a function of the 3-dimensional coordinate $\underline{x}$. The conservation equation of energy leads in the presence of a heat source term, $p(\underline{x})$, and by introducing the heat capacity, $c(\underline{x})$, to the dynamic differential equation:

$$c(\underline{x}) \frac{\partial}{\partial t} T(\underline{x},t) - \nabla(\kappa(\underline{x},T(\underline{x},t)) \cdot \nabla T(\underline{x},t)) = p(\underline{x}) \tag{3.2}$$

For the static case, the differential equation:

$$-\nabla(\kappa(\underline{x},T(\underline{x})) \cdot \nabla T(\underline{x})) = p(\underline{x}) \tag{3.3}$$

is valid.

Membrane-like microstructures are generally several micrometers thick, while the lateral dimensions of the structures and the surrounding package are on the order of a few hundred micrometers. If the layered thin-film structure would be directly transferred to a 3-d geometry model, an enormous number of finite elements would be created, as the smallest structure size determines the mesh density. Averaging the structural information and properties over the different layers in the cross section of the membrane is a good method to avoid such problems. The membrane is, therefore, initially treated as a quasi-two-dimensional object.

To account for the much larger lateral extension of the membrane in comparison to its thickness, some approximations are made. The variation of the temperature field in the membrane along the z-axis is assumed to be small, i.e., the temperature is nearly constant. For further considerations, the temperature field within the membrane is separated into two terms:

$$T\left(\underline{x},t\right) = \tilde{T}(z) \cdot T(x,y,t)|_{z=0} \tag{3.4}$$

where the lateral membrane temperature field $T(x,y,t)$ is modulated in the z direction by the function $\tilde{T}(z)$. The coordinate $z = 0$ corresponds to the top surface of the membrane. This function is developed in a power series:

$$\tilde{T}(z) = \sum_{n=0} a_n z^n \tag{3.5}$$

The dynamic Eq. (3.2) is integrated along the z-axis within the interval of the topmost membrane layer at $z = 0$ and the local membrane thickness z_d:

$$\int_{-z_d}^{z=0} c\left(\underline{x}\right) \frac{\partial}{\partial t} T\left(\underline{x},t\right) dz - \int_{-z_d}^{z=0} \nabla\left(\kappa\left(\underline{x},T\right) \cdot \nabla T\left(\underline{x},t\right)\right) dz = \int_{-z_d}^{z=0} p\left(\underline{x}\right) dz \tag{3.6}$$

The first integral can be rewritten with Eq. (3.4) as:

$$\int_{-z_d}^{z=0} c\left(\underline{x}\right) \frac{\partial}{\partial t} T\left(\underline{x},t\right) dz = \left(\int_{-z_d}^{z=0} c\left(\underline{x}\right) \tilde{T}(z) dz\right) \cdot \frac{\partial}{\partial t} T(x,y,t)$$

$$= c_{\mathrm{m}}(x,y) \cdot \frac{\partial}{\partial t} T(x,y,t) \tag{3.7}$$

The second term in Eq. (3.6) is divided into two parts by using a 2-dimensional gradient:

$$\int_{-z_d}^{z=0} \nabla\left(\kappa\left(\underline{x},T\right) \cdot \nabla T\left(\underline{x},t\right)\right) dz = \nabla_{xy}\left(\left(\int_{-z_d}^{z=0} \kappa\left(\underline{x},T\right) \tilde{T}(z) dz\right) \nabla_{xy} T(x,y,t)\right)$$

$$+ T(x,y,t) \cdot \left(\int_{-z_d}^{z=0} \frac{\partial}{\partial z}\left(\kappa\left(\underline{x},T\right) \frac{\partial}{\partial z} \tilde{T}(z)\right) dz\right) \tag{3.8}$$

The first term on the right-hand side of Eq. (3.8) corresponds to a two-dimensional heat flux. The second part has, in this model, the physical interpretation of a local heat sink or source perpendicular to the x-y-plane. The equation can be further simplified by introducing an integral heat conductivity within the membrane:

$$\kappa_{\mathrm{m}}(x,y,T) = \int_{-z_d}^{z=0} \kappa\left(\underline{x},T\right) \tilde{T}(z) dz \tag{3.9}$$

The power generation term on the right-hand side of Eq. (3.6) becomes:

$$\int_{-z_d}^{z=0} p\left(\underline{x}\right) dz = p_{\mathrm{m}}(x,y) \tag{3.10}$$

As already mentioned, the thickness of the membrane is much smaller than the lateral dimensions. This motivates the assumption that the temperature across the membrane is approximately constant in the z-direction, which modifies Eq. (3.5) to:

$$\tilde{T}(z) = a_0 = 1 \tag{3.11}$$

Rewriting the dynamic differential Eq. (3.1) with the results of Eqs. (3.7), (3.8) and (3.9), yields:

$$c_m(x,y)\frac{\partial}{\partial t}T(x,y,t) - \nabla_{xy}\left(\kappa_m(x,y,T)\cdot\nabla_{xy}T(x,y)\right) = p_m(x,y) \tag{3.12}$$

Since the relation

$$\frac{\partial}{\partial z}\tilde{T}(z) = 0 \tag{3.13}$$

holds, the second integral on the right-hand side of Eq. (3.8) vanishes.

The membrane consists of a stack of thin films of a certain thickness, d_i, with characteristic isotropic and homogeneous material constants c_i and κ_i (see Fig. 3.2). The composition of the stack locally depends on the x,y-coordinates. The local membrane thickness is defined as

$$d_m(x,y) = \sum_i d_i \tag{3.14}$$

and varies according to the microhotplate structure.

Integration of the material constants over z leads to:

$$c_m(x,y) = \int_{-z_d}^{z=0} c(\underline{x})\tilde{T}(z)\,dz = \sum_i c_i d_i \tag{3.15}$$

$$\kappa_m(x,y,T) = \int_{-z_d}^{z=0} \kappa(\underline{x},T(\underline{x},t))\tilde{T}(z)\,dz = \sum_i \kappa_i d_i \tag{3.16}$$

The temperature dependence of the heat-conductivity for a certain film can be included as a truncated Taylor series:

$$\kappa_i(T) = \kappa_{i0}\left(1 + \beta_i\Delta T\right) \tag{3.17}$$

where

$$\kappa_{i0} = \kappa_i(T_0) \tag{3.18}$$

is the heat conductivity at the reference temperature T_0, e.g., ambient temperature, corresponding to the unheated state of the microhotplate. β_i is the first-order temperature coefficient of the heat conductivity and $\Delta T = T - T_0$.

The heat generation and loss term $p_m(x,y)$, can be subdivided into several contributions:

$$p_m(x,y) = p_{Joule} - p_{rad} - p_{air} \tag{3.19}$$

The heat on the hotplate is generated by resistive Joule heating, p_{Joule}. Usually, one resistive layer in the membrane serves as heating structure, while all others do not contribute to heating. With the thickness of the heating layer, d_{heat}, the Joule-heating can be expressed as:

$$p_{\text{Joule}} = \sigma(T) \cdot \nabla V^2 \cdot d_{\text{heat}} \qquad (3.20)$$

A possible heat-loss mechanism includes thermal radiation, p_{rad}. The hotplate operating temperature range is up to 350 °C, for which radiation losses are considered to be negligible [94, 96]. In case of higher temperatures, radiation losses would have to be included [97, 98]. The overall loss owing to radiation scales with the total heated area. A rough estimate for radiation losses of the presented microhotplate at 300 °C is 2% of the overall hotplate power consumption.

On the other hand, there are heat losses through the surrounding air, p_{air}. These can be further subdivided into pure conduction losses owing to the heat conductivity of the air, natural convection losses and forced-convection losses. In the present case, natural convection can be neglected [99], which was validated by measuring the power consumption of a certain membrane at different orientations in air. The power consumption did not depend on the orientation such as the inclination angle of the hotplate, which would have been the case if natural convection had played an important role. A further contribution to the heat losses through air is forced convection originating from streaming air, which also can be neglected in the model owing to the encapsulation of the sensor. Consequently only the heat losses owing to the heat conduction of the air remain to be considered.

To correctly assess the contribution of p_{air}, it is necessary to transform the quasi-2-d model of the membrane back into a 3-d geometric FEM model, as it is schematically shown in Fig. 3.2. The membrane is modeled as a plate of defined thickness, d_{mod}. A slightly larger d_{mod} in comparison to the physical thickness also increases the size of the mesh units in the membrane and leads to a reduced total number of elements without significantly affecting the modelling results. The local material properties in the membrane, $c_{\text{m}}(x,y)$, $\kappa_{\text{m}}(x,y)$, and the heat generation, p_{Joule}, are converted back into volumetric units by division with d_{mod}. With this method, the topographical and material thin film structure is included in the averaged, homogeneous material properties of the membrane model that features a constant membrane thickness.

3.3 FEM-Simulations

FEMLAB™ is a MATLAB™-based finite-element program with direct access to the model equations. The static nonlinear heat conductivity mode was chosen for the simulation, in which the nonlinear coefficients are implemented through polynomials representing the temperature field. Starting with the physical layout, a geometrical model for the solver was constructed as was shown in Fig. 3.2 and was described in the previous section. The designs presented in the next chapter are intended to feature homogeneous temperature distribution and low stress gradients. The symmetry is advantageous for modelling, since it is sufficient to model only a representative fraction (one quarter) of the structure. The total width of metal lines connecting the functional elements on the heated area is summed up and then divided by four. As already mentioned, the integrated heat conductivity, $\kappa_m(x,y,T)$, is divided by the modelling thickness, d_{mod}. A modelling thickness of the membrane, d_{mod}, of 7.5 μm was used which is close to the averaged membrane thickness. Since d_{mod} represents a geometrical model simplification and is not the physical thickness of the membrane, its value can be chosen in a certain range. This value determines the minimum mesh size and, therefore, the total number of elements created by the meshing tool. The chosen value of d_{mod} allows for computing a solution at reasonable computational efforts.

The heat is generated in the resistive microhotplate heater. The power density was calculated for the total volume of the heater. The surfaces were set to T_0, so that the temperature increase upon heating was modelled, and the coefficients refer to ambient temperature ($T_0 = 25\,°C$). The heat conductivity values and temperature coefficients of the CMOS layers have been measured for the CMOS process family, which is used to fabricate the devices [103]. The doping level in the n-well silicon island does not significantly alter the heat conductivity in the temperature range of interest so that the value of single-crystalline silicon was used [105, 106]. The silicon island size and the membrane size vary owing to wafer-thickness and etching-process fluctuations. These fluctuations also influence the measured thermal resistance, which makes it imperative to use the real geometric values of fabricated devices for validating the model. The heat conductivity of the surrounding air was modeled on the basis of measured values [107] leading to the function: $k_{air}(\Delta T) = 26.3 + 0.079\Delta T - 3.38 \cdot 10^{-5}\Delta T^2$. It is important to point out that no fitting or open parameters were used in the entire simulation. The mesh was generated automatically with the meshing tools. The values of the heat conductivities and heat capacities, which were used for the simulation, are summarized in Table 4.2, the respective results will be presented and compared to the measured data in Chap. 4.

3.4 Lumped Microhotplate Model

In the previous paragraph, the basic considerations of FEM modelling have been laid out. The outcome of a static thermal simulation based on this model is a 3-d temperature field $T(x,y,z)$. In this section it is discussed, how the characteristic figures, such as thermal resistance and thermal time constant of the membrane, can be deduced.

The temperature sensor on the hotplate measures the membrane temperature, T_M, at a certain location, $\underline{x}_s$, such as the membrane center:

$$T_M = T\left(\underline{x}_s\right) \tag{3.21}$$

If the sensor occupies a larger area, the temperature field has to be averaged accordingly. Another characteristic number is the overall power consumption, P_{heat}, at a certain temperature distribution. The Joule heating, as discussed in Eq. (3.20), requires a complex electrothermal simulation. Such simulations can be circumvented by introducing a local heat generation density in the membrane under the prerequisite that the heated area shows a homogeneous temperature distribution, and that no hot spots appear as a consequence of heater resistance variations. Otherwise, the heat generation distribution would be affected, since electrical conductivity, and, consequently the heater resistance, exhibit significant temperature dependence. The conservation of energy requires that the total heat flux through the outer boundary, $\partial\Omega$, of the structure is equal to the total power generation, P_{heat}:

$$P_{heat} = \oint_{\partial\Omega} j \cdot dA = V_{heat} \cdot I_{heat} \tag{3.22}$$

On the other hand, the total heating power can be calculated as the product of heating voltage, V_{heat}, and heating current, I_{heat}. For a heater occupying an area, A_{heat}, in the geometric model, the power density can be calculated to:

$$P_{Joule} = \frac{P_{heat}}{A_{heat} \cdot a_{mod}} \tag{3.23}$$

A characteristic measurement includes the determination of the microhotplate temperature as a function of its power consumption. The curves can be fitted by a second-order polynomial using the coefficients η_0 and η_1,

$$\Delta T = T_M - T_0 = \eta_0 P_{heat} + \eta_1 P_{heat}^2 \tag{3.24}$$

The reference temperature, T_0, refers to ambient temperature, i.e., the membrane temperature before applying any heating power. The thermal resistance, η, can be generally defined as:

$$\eta(T) = \left. \frac{dT_M}{dP_{heat}} \right|_T \tag{3.25}$$

The thermal resistance will be temperature-dependent as can be seen in Eq. (3.24), which is not only a consequence of the temperature dependence of the thermal heat conduction coefficients. The measured membrane temperature, T_M, is related to the location of the temperature sensor, so that the temperature distribution across the heated area will also influence the thermal resistance value. The nonlinearity in Eq. (3.24) is, nevertheless, small. The expression "thermal resistance" consequently often refers to the coefficient η_0 only, which is used as a figure of merit and corresponds, according to Eqs. (3.24) and (3.25), to the thermal resistance or thermal efficiency of the microhotplate at ambient temperature, T_0. The temperature T_M can be determined from simulations with distinct heating powers. The thermal resistance then can be extracted from these data.

Another value that can be calculated from the simulation results, is the total amount of stored heat:

$$\Delta Q = \int_V c(\underline{x}) (T(\underline{x}) - T_0) \, dV \tag{3.26}$$

Using a linear approximation leads to:

$$c_0 = \frac{\Delta Q}{\Delta T} \cong \left. \frac{dQ}{dT_M} \right|_{T_0} \tag{3.27}$$

where c_0 denotes the total heat capacitance of the microhotplate at the reference temperature, T_0.

Using the thermal resistance and the total heat capacitance, the dynamic equation for a lumped-element model in the linear regime can be written as:

$$c_0 \cdot \frac{\mathrm{d}}{\mathrm{d}t}\Delta T + \frac{1}{\eta_0}\Delta T = P_{\text{heat}} \qquad (3.28)$$

the solution as:

$$T(t) = \Delta T_{\text{f}}\big(1 - \exp(-t/\tau)\big) \qquad (3.29)$$

with

$$\Delta T_{\text{f}} = T_{\text{f}} - T_{\text{i}} \qquad (3.30)$$

where T_{f} denotes the final temperature and T_{i} the initial temperature.

For a heating process starting at ambient temperature, Eq. (3.28) for the static case can be written as:

$$\Delta T_{\text{f}} = T_{\text{f}} - T_0 = \eta_0 P_{\text{heat}} \qquad (3.31)$$

which is equivalent to a linearized version of Eq. (3.24). The time constant in Eq. (3.29) can be expressed as:

$$\tau_0 = c_0 \eta_0 \qquad (3.32)$$

which can be determined by measuring the temporal behavior of the microhotplate temperature. The experimental time constant is calculated from a dynamic measurement curve using a fit based on Eq. (3.29). On the other hand, the simulated time constant can be deduced from the result of the static simulation. Inserting Eq. (3.27) and Eq. (3.31) in Eq. (3.32) leads to:

$$\tau_0 = \frac{\Delta Q}{\Delta T}\frac{\Delta T}{P_{\text{heat}}} = \frac{\Delta Q}{P_{\text{heat}}} \qquad (3.33)$$

The time constant will show a temperature dependence owing to thermal resistance variation and temperature-dependent heat capacitance. The differential – and more general – form of Eq. (3.33) is:

$$\tau(T) = \frac{\mathrm{d}Q}{\mathrm{d}P_{\text{heat}}}\bigg|_T \qquad (3.34)$$

The considerations so far rely on constant heating power, and the way how this power is applied to the microhotplate does not play a role. In fact, a monolithically integrated control circuitry does not apply constant power but acts as an adjustable current source. Moreover, for measuring the thermal time constant experimentally, either a rectangular voltage or rectangular current pulse is applied. Analyzing the dynamic temperature response of the system leads to a measured time constant, which differs from the "real" thermal time constant as defined by Eqs. (3.33) and (3.34). The reason is the "self-heating" in the resistive heater, which will be exemplified for the case of a current-source driven heater. In a first-order approximation, the heating power of such a heater that is fully located on a membrane can be expressed as:

$$P_{\text{heat}} = R_{\text{heat}} \cdot I_{\text{heat}}^2 = R_0 \left(1 + \alpha_{\text{h}} \cdot \Delta T\right) \cdot I_{\text{heat}}^2 \tag{3.35}$$

In case of a homogeneous temperature distribution in the heated area, α_{h} corresponds to the temperature coefficient of the heater material, otherwise α_{h} includes the effects of temperature gradients on the hotplate. As a consequence of the already mentioned self-heating, the applied power is not constant over time, and the hotplate cannot be simply modelled using a thermal resistance and capacitance. Replacing the right-hand term in Eq. (3.28) by Eq. (3.35) leads to a new dynamic equation:

$$c_0 \cdot \frac{\text{d}}{\text{d}t} \Delta T + \left(\frac{1}{\eta_0} - R_0 \cdot \alpha_{\text{h}} \cdot I_{\text{heat}}^2\right) \Delta T = R_0 \cdot I_{\text{heat}}^2 \tag{3.36}$$

The coupling of the hotplate to the electronics via a resistive heater consequently alters the equation, and a change in the effective time constant occurs:

$$\tau_{\text{eff}} = c_0 \eta_{\text{eff}} = \tau_0 \cdot \left(1 - \eta_0 R_0 \cdot \alpha_{\text{h}} \cdot I_{\text{heat}}^2\right)^{-1} \tag{3.37}$$

Inclusion of the self-heating effect yields an additional temperature dependence of the thermal time constant. Differences in the time constants for heating and cooling are evident, and the real thermal time constant can be observed only in the cooling cycle with $I_{\text{heat}} = 0$.

3.5 AHDL-Model for System Simulations

The modeling of micro-electro-mechanical systems (MEMS) with hardware description languages (HDLs) supports the design and simulation of microsystems, which include MEMS-components and readout electronics [108]. For a complete and reliable simulation of the micromechanical sensor system it is mandatory to find a model for the microhotplate in an analog hardware-description language (AHDL). AHDL is a computer language that is specifically designed to enable high-level description and the behavioral simulation of continuous-time systems.

Examples for such descriptions were reported recently. Starting with the lumped-model equations explained in Sect. 3.4, a matrix formulation can be found that supports the system optimization [19].

A first description of the microhotplate in AHDL was developed, which calculates the power dissipated by the polysilicon heater as shown in Fig. 3.3 [89]. The calculated power serves as input for a look-up table with the measured values of the power dissipated by a normalized polysilicon resistor, which then provides the corresponding microhotplate temperature. The model extracts the microhotplate temperature from the table. This microhotplate temperature is subject to temporal delay

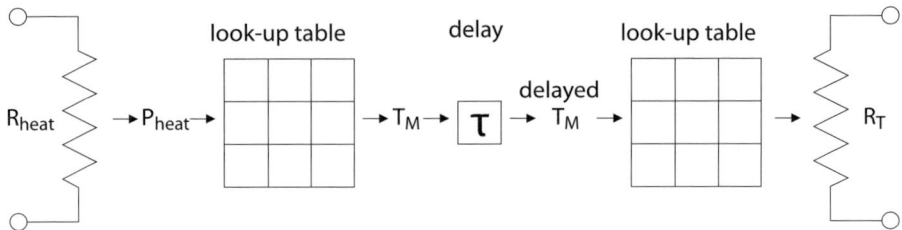

Fig. 3.3. AHDL description of a microhotplate using look-up tables

according to the thermal time constant of the microhotplate. The microhotplate temperature then serves as input to another look-up table containing the measured values of microhotplate temperatures and the corresponding resistance values of a normalized polysilicon resistor. The model thereafter extracts the resistance value from the table to calculate the resistance of the polysilicon temperature sensor.

This table can be replaced by analytic expressions, which reduces the input parameter set to variables that can be determined by simulations or measurements on microhotplate test devices. The input signal is the heating power applied to the microhotplate:

$$P_{\text{heat}}(t) = I_{\text{heat}}(t) \cdot V_{\text{heat}}(t) \tag{3.38}$$

According to Eq. (3.31), the heating power would produce a final static temperature difference, ΔT_f. Now the dynamics have to be taken into account. A reformulation of the differential equation leads to an expression including a certain delay:

$$\Delta T(t) = \Delta T_f - \tau \frac{dT}{dt} \tag{3.39}$$

As already mentioned, the applied power to the hotplate is not constant owing to the change of the heater resistance:

$$R_{\text{heat}}(t) = f(\Delta T(t)) = R_{\text{heat}}(T_0) \cdot (1 + \alpha_h \Delta T(t)) \tag{3.40}$$

The self-heating effect of the heater is included in a change of the heating voltage:

$$V_{\text{heat}}(t) = R_{\text{heat}}(t) \cdot I_{\text{heat}}(t) \tag{3.41}$$

In order to allow for a complete analysis of the sensor device, the feedback provided by the temperature sensor on the membrane has to be included as well. The voltage across the temperature sensor resistor can be determined as:

$$V_T(t) = I_{\text{bias}} \cdot R_T(\Delta T(t)) \tag{3.42}$$

In the simulation, these equations describing the hotplate are coupled to the circuitry. The voltage, $V_T(t)$, determines the output current, $I_{\text{heat}}(t)$, of the circuitry, but the circuitry response time is much smaller than the thermal time constant of the microhotplate. A coupled set of equations, those of the system including the microhotplate and the circuitry, is solved during the overall simulation procedure.

4

Microhotplates in CMOS Technology

The following chapter includes the description of different types of microhotplates that feature resistor and transistor heating elements. Three of them were specifically designed to be monolithically integrated with circuitry, and one was a testing device that was used for the assessment of temperature distributions on the microhotplates.

The first device is a circular microhotplate (Sect. 4.1). One important guideline was to implement the microhotplate in CMOS technology with a minimum of post-CMOS micromachining steps. Additionally the hotplate had to be optimized for drop-coating with nanocrystalline tin-oxide layers. This microhotplate was co-integrated with circuitry, and the respective monolithic sensor system will be discussed in Sect. 5.1.

The second microhotplate design is derived from this circular microhotplate. In contrast to the first device, it does not feature a silicon island underneath the heated area, but exhibits a network of temperature sensors in order to assess the temperature distribution and homogeneity (Sect. 4.2). The measured temperature distribution was compared to simulations, and the model described in Chap. 3 was validated. Furthermore, the influence of the tin-oxide droplet on the temperature distribution was studied. A microhotplate without silicon island is much easier to fabricate, though the issue of sufficient temperature homogeneity has to be evaluated.

The third microhotplate introduced in Sect. 4.3 was designed to extend the operation temperature limit imposed by the CMOS-metallization contacts in the heated area. A new heater design was devised, and a microfabrication sequence that enables the realization of Pt temperature sensors and Pt-electrodes was developed. This microhotplate was also monolithically integrated with circuitry as presented in Sect. 5.2, and operating temperatures of up to 500 °C have been achieved.

Finally, a transistor-heated hotplate will be described (Sect. 4.4), which offers the advantage of lower power consumption, since there is no additional power transistor needed on the chip. Moreover, a transistor hotplate can be digitally controlled and addressed so that new operation modes can be realized (Sect. 4.5). The integration of the transistor hotplate with accompanying, mostly digital circuitry will be described in Sect. 6.3.

4.1 Circular Microhotplate

4.1.1 Design Considerations

The microhotplate design and development was guided by the following considerations:

– The membrane layout should be as symmetric as possible to achieve good temperature homogeneity over the membrane area and, as a consequence, low stress gradients. This includes also thermal stress owing to the mismatch of the thermal expansion coefficients of the layer materials.
– A homogeneous temperature distribution in the heated area is highly desirable to make sure that all sensing processes on the hotplate take place at the same defined and precisely controlled temperature.
– A high thermal resistance is needed to achieve minimum power consumption.
– The desired operating temperature (250 °C–350 °C) has to be reached with a supply voltage of 5.5 V.

In view of the above considerations, a circular design of the heated area (300 μm diameter) on a square dielectric membrane (500 × 500 μm^2) was chosen, which is shown in Fig. 4.1.

The design parameters of this microhotplate are summarized in Table 4.1.

A considerable part of the heat is usually dissipated through the metal supply lines to the bulk. A circular-shape microhotplate reduces this heat dissipation, as it allows for comparatively long metal leads extending from the heated area to the corners of the membrane. The width of the metal lines along each diagonal is approximately identical. Thus, a symmetric heat flow through the membrane is created. The heat is produced by a ring heater along the edge of the circular heated area.

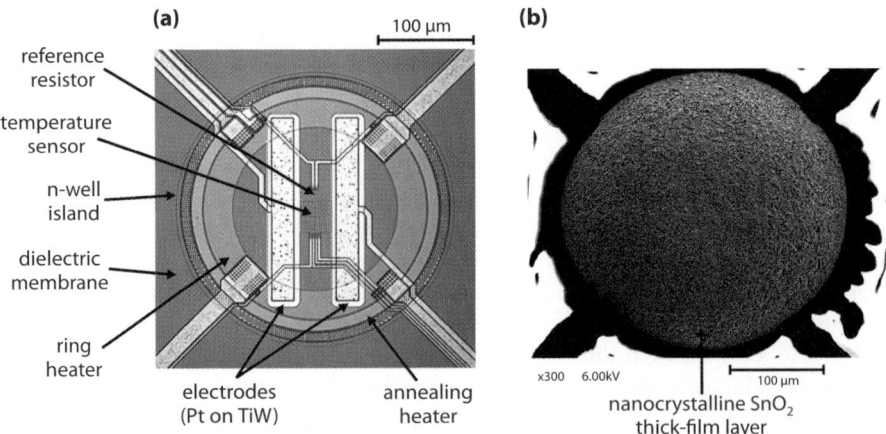

Fig. 4.1. (a) Micrograph of a circular microhotplate without sensitive layer, (b) SEM-micrograph of a metal-oxide coated microhotplate

Table 4.1. Design parameters of the circular microhotplate

membrane size	$500 \times 500\,\mu m^2$
Si-island diameter	$300\,\mu m$
heater resistance	$125\,\Omega$
annealing heater resistance	$600\,\Omega$
temperature sensor resistance	$10\,k\Omega$
reference resistor	$3\,k\Omega$
electrode distance	$40\,\mu m$
electrode length	$205\,\mu m$

The circular heater design also perfectly matches the shape of the sensitive tin-oxide droplet: No excess area is heated, and the heat losses to ambient air are reduced. A SEM (Scanning Electron Microscope) micrograph of a microhotplate coated with a SnO_2-droplet is shown in Fig. 4.1. The grainy structure of the nanocrystalline oxide material is clearly visible.

From the conventional heat conduction equation it is found that the temperature gradient within the heated area can be minimized (a) by using high-thermal-conductivity materials in the heated area and (b) by generating the heat at the location with the largest thermal losses. Both issues were taken into account in realizing the hotplate heated-area design. First, a 5.5-μm-thick n-well silicon island, which effectively distributes the heat, owing to the high thermal conductivity of silicon, was fabricated underneath the heated area in the center of the membrane. The island also mechanically stabilizes the membrane, which results in less membrane buckling. The second feature is a novel type of polysilicon ring heater including two semicircular heating resistors along the edges of the heated area, which are connected in parallel. Using this heater type, the heat is generated in the locations with the largest heat losses via membrane and metal leads to the bulk. The parallel heater configuration moreover decreases the overall heater resistance to a nominal value of $125\,\Omega$. With such a low resistance it is possible to provide enough heating power to reach high temperatures (up to $350\,^\circ C$) with the on-chip driving circuitry at $5.5\,V$ supply voltage.

The temperature sensor in the membrane center is made of polysilicon with a nominal resistance of $10\,k\Omega$. An additional reference resistor is needed for the control circuitry (Sect. 5.1). For the resistance measurement of the sensitive layer, platinum electrodes are deposited on top of the CMOS aluminum metallization in order to establish good electrical contact to the sensitive metal oxide.

An additional heater for annealing of the sensitive material was also integrated but not used so far. The basic idea was to provide a heater, which can achieve much higher temperatures than the heater for normal operation. Since the temperature for material sintering on wafer-level is limited to $400\,^\circ C$ to prevent degradation of the integrated electronics (last thermal step of the CMOS process is $400\,^\circ C$), higher annealing temperatures can only be applied locally on the membranes by using the extra heater with an external power source. The annealing heater is, if ever, only used once during sensor fabrication. Therefore, only the heater for normal sensor operation is connected to the circuitry in the monolithic sensor system (see Sect. 5.1) and was optimized for performance.

4.1.2 Fabrication

The starting material for the sensor fabrication are fully processed wafers of a 2-poly 2-metal 0.8 μm industrial CMOS process provided by austriamicrosystems (Unterpremstätten, Austria). In the following, the main post-CMOS processing steps (schematically summarized in Fig. 4.2) are discussed.

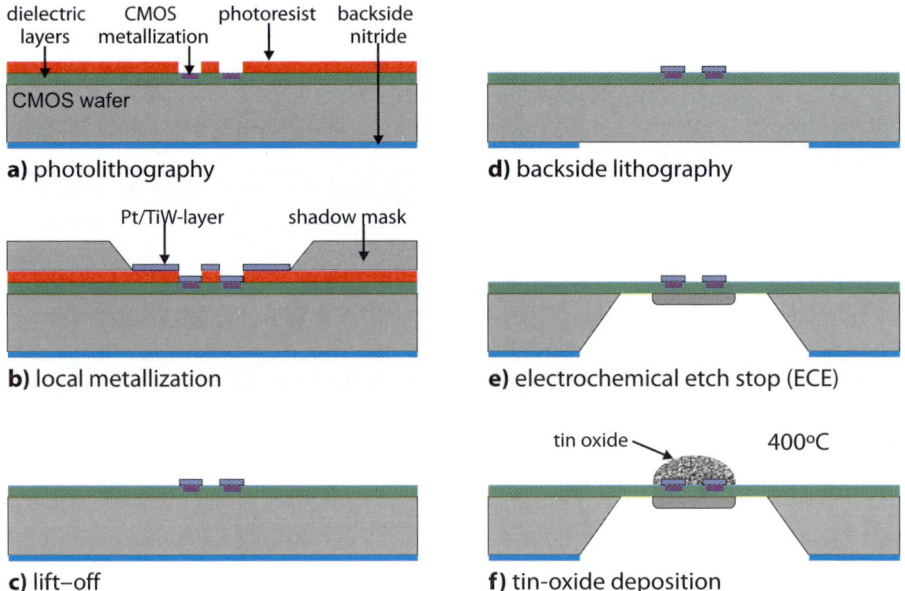

Fig. 4.2. Schematic of the process flow to fabricate the circular microhotplate

Local Metallization

Good electrical contact to the sensitive layer is required for optimal sensor operation. The CMOS aluminum metallization turned out not to be well suited for contacting the tin-oxide droplet. Oxidation of the coated electrodes was observed during annealing. As a consequence, the contact resistance increased, and no sensor signal was measurable. To circumvent this problem, a lift-off metallization process for covering the electrodes with a better contact metal, such as platinum, was developed. Lift-off techniques are well-suited for post-processing as no aggressive chemistry for patterning is involved. Moreover, to assure good lift-off results with clean edges, a negative-undercut resist profile is desirable. Standard profiles suffer from metal deposition on the side walls, which remain as residues after resist removal. One possible solution is a process involving an image-reversal resist.

After a cleaning sequence including an O_2-plasma step, the photolithography is performed (Fig. 4.2a) using the image-reversal resist TI35ES (microchemicals, Berlin, Germany). A dark-field-structured quartz mask is applied to transfer the desired structures into the resist. After a reversal bake, a flood exposure is performed, and only the structures, which have been exposed, are dissolved in the developer AZ826MIF (Clariant). This developer is particularly well-suited for a lift-off process, since it completely removes resist residues by special additives. After the development, the desired undercut resist profile was achieved, as has been established by using SEM (Fig. 4.3).

2.5 µm

Fig. 4.3. SEM micrograph of image-reversal resist stripes of a test pattern after photolithography. The resist shows an undercut profile as desired for the lift-off process

A shadow-mask technique has been applied for the local metal deposition to exclude metal residues on other designs processed on the same wafer (Fig. 4.2b). Such metal residues may be caused by imperfections in the patterned resist due to topographical features on the processed CMOS wafers or dust particles. The metal film is only deposited in those areas on the wafer, where it is needed for electrode coverage on the microhotplates. This also renders the lift-off process easier since no closed metal film is formed on the wafer, so that the acetone has a large surface to attack the photoresist. Another advantage of the local metal lift-off process is its full compatibility with the fabrication sequence of chemical sensors based on other transducer principles [20].

A silicon wafer with anisotropically KOH-etched openings was used as shadow mask. The shadow mask is accurately positioned with the help of an optical microscope and fixed using a custom-made wafer holder. A 50-nm-thick TiW-film is deposited by sputtering through the shadow mask. This film serves as adhesion layer and diffusion barrier and covers the rough surface of the CMOS-Al-metallization. A Pt-layer with a thickness of 100 nm was sputtered on top of this TiW-layer.

Finally, the wafer is immersed in acetone, and the lift-off process is performed in an ultrasonic bath. The Pt metal layer showed good adhesion to the CMOS-aluminum electrodes and facilitated good electrical contact to the sensitive layer.

Membrane Release

The membranes of the microhotplates were released by anisotropic, wet-chemical etching in KOH. In order to fabricate defined Si-islands that serve as heat spreaders of the microhotplate, an electrochemical etch stop (ECE) technique using a 4-electrode configuration was applied [109]. ECE on fully processed CMOS wafers requires, that all reticles on the wafers are electrically interconnected to provide distributed biasing to the n-well regions and the substrate from two contact pads [110]. The formation of the contact pads and the reticle interconnection requires a special photolithographic process flow in the CMOS process, but no additional non-standard processes.

Another CMOS-process modification included the deposition of a nitride layer on the wafer backside. The backside nitride is identical with the CMOS passivation. All wafers are already delivered with this backside nitride by the CMOS foundry austria*micro*systems.

After the lift-off process, the nitride is structured in a RIE etching process with SF_6 to define the etch windows for the membrane openings on the backside (Fig. 4.2d). Afterwards, a protective coating is spin-coated onto the frontside to protect the open CMOS metallization features and to stabilize the fragile membrane structures. The wafer is then mounted in a waferholder and connected to a potentiostat (AMMT, Frankental, Germany). Finally the wafer in the holder is immersed in 6M KOH solution at 90 °C, and the membranes are released (Fig. 4.2e). After etching, the protective coating is completely removed in a O_2-plasma microstripper.

Dicing

Conventional chip-dicing saws apply high water pressures to remove the debris from the chip surfaces during dicing. As the released membranes would not survive such a standard dicing process, the chips were diced applying the lowest possible water pressure. Additionally, the wafer frontside was protected using an adhesive foil (Adwill P-5780, Lintec Corporation, Japan), which easily peels off after UV-illumination and heating. The foil stabilized the membranes and protected them against the deposition of sawing dust [111]. The optimized lift-off, etching and dicing process led to an excellent fabrication yield of 95% for the microhotplates.

Deposition of the Sensitive Layer

The basics of the paste preparation were explained in Sect. 2.3.3. For the devices presented in this book, the paste was deposited onto cleaned chips using a drop-coating method [48, 61]. The deposition was performed by the company Applied-Sensor (Reutlingen, Germany). A metal-wire loop is immersed in the paste and the tin-oxide suspension adhering to the loop forms a droplet, which is accurately positioned in the membrane center. After the drop deposition, the whole chip is put in a belt oven and annealed for 20 min at a temperature of 400 °C. This temperature is close to the elevated-temperature steps at the backend of the CMOS process. Consequently, we never observed a significant difference of the circuitry performance between coated and uncoated chips. The whole deposition process is, therefore, fully CMOS compatible, and no additional on-chip annealing is necessary.

The chips have not to be bonded before the layer deposition, and the droplet deposition and annealing can be performed on wafer level, which is advantageous for device commercialisation.

To achieve higher annealing temperatures, the additional on-chip annealing heater was implemented. In-situ annealing requires the chip to be packaged and wire-bonded prior to processing the sensitive material. Only single-chip processing is possible. Therefore, the additional annealing heater was never used so far. The annealing temperature range can be extended up to 500 °C by using the hotplate annealing heater.

4.1.3 Physical Microhotplate Characterization

For thermal characterization and temperature sensor calibration a microhotplate was fabricated, which is identical to that on the monolithic sensor chips, but does not include any electronics. The functional elements of this microhotplate are connected to bonding pads and not wired up to any circuitry, so that the direct access to the hotplate components without electronics interference is ensured. The assessment of characteristic microhotplate properties, such as the thermal resistance of the microhotplate and its thermal time constant, were carried out with these discrete microhotplates.

A schematic view of the microhotplate functional elements is presented in Fig. 4.4. A resistive temperature sensor is embedded in the heated area of the microhotplate. The resistance is measured in a four-point measurement. The calibration procedure of the temperature sensor will be explained in the next section (Sect. 4.1.4). The heating power dissipation is determined using also a four-point configuration. The external wiring of the heater typically adds another 5% to the heater resistance, which has to be eliminated for an accurate measurement of the dissipated power. A heating current, I_{heat}, is applied, and the voltage drop, V_{heat}, across the heater is measured on chip.

The setup includes multimeters and current sources controlled by a LABVIEW™ program. Equidistant power steps can be generated by the program, which is advantageous for extracting thermal resistances. The required heating current is ramped up and down. This ramping allows for detecting possible hysteresis effects in the temperature sensor measurements and in the chip heating, none of which has been found in

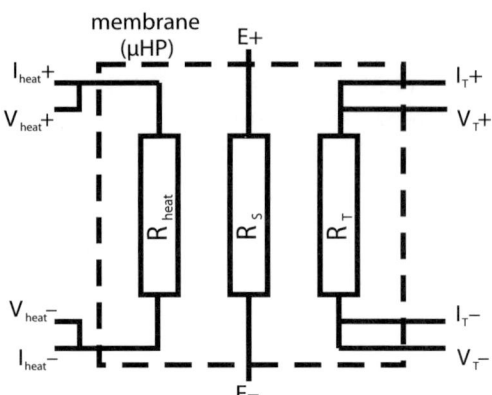

Fig. 4.4. Schematic of the functional elements of the microhotplate. R_{heat} denotes the heating resistance, R_S is the metal-oxide resistance measured between the electrode contacts $E+$ and $E-$, and R_T is the temperature-sensor resistance

the measurements. Using a calibration polynomial for each sensor (see Sect. 4.1.4), the temperature rise upon membrane heating was recorded, and the resulting correlation between heating power and microhotplate temperature was established.

In order to determine the thermal time constant of the microhotplate in dynamic measurements, a square-shape voltage pulse was applied to the heater. The pulse frequency was 5 Hz for uncoated and 2.5 Hz for coated membranes. The amplitude of the pulse was adjusted to produce a temperature rise of 50 °C. The temperature sensor was fed from a constant-current source, and the voltage drop across the temperature sensor was amplified with an operational amplifier. The dynamic response of the temperature sensor was recorded by an oscilloscope. The thermal time constant was calculated from these data with a curve fit using Eq. (3.29). As already mentioned in the context of Eq. (3.37), self-heating occurs with a resistive heater, so that the thermal time constant has to be determined during the cooling cycle.

4.1.4 Calibration of the Temperature Sensors

The discrete microhotplates were packaged and bonded in a DIL-28 package for temperature sensor calibration. A Pt-100-temperature sensor was attached to the chip package in close vicinity to the sensors. The chips were then calibrated in an oven at temperatures up to 325 °C with the help of the Pt-100 resistor. A second-order polynomial was extracted from the measurements for each temperature sensor providing the temperature coefficients α_1 and α_2:

$$\Delta T = T - T_0 = \alpha_1 \rho + \alpha_2 \rho^2 \tag{4.1}$$

where ρ is the relative resistance change and $R(T)$ the measured resistance of the temperature sensor at temperature T:

$$\rho = \frac{R(T) - R(T_0)}{R(T_0)} \tag{4.2}$$

The calibration was done for temperatures between 25 °C and 325 °C. The data were fitted according to Eq. (4.1), and the two temperature coefficients were extracted. Within the same wafer batch, a production-spread-induced error of 2% in the determination of the microhotplate temperature coefficients was observed.

4.1.5 Comparison of Thermal Characterization and Simulation Results

The circular microhotplate was thermally characterized, and the results were compared with simulations carried out according to the approach discussed in Chap. 3. Applying FEM simulations as described in Sect. 3.3 generate a temperature field, and the temperature in the membrane center represents the overall membrane temperature according to Eq. (3.21). The values that have been used for the simulation are summarized in Table 4.2.

The relationship between the temperature difference, ΔT, and the input power is shown in Fig. 4.5 for microhotplate simulations and measurements. The simulated values are plotted together with the mean value of the experimental data for a set of three hotplates of the same wafer. The experimental curve was fitted with a second-order polynomial according to Eq. (3.24). As a result of the curve fit, the thermal resistance at room temperature, η_0, is 5.8 °C/mW with a standard deviation of ±0.2 °C/mW, which is mainly due to variations in the etching process.

A thermal resistance of 5.7 °C/mW results from the simulation in the range between 0 and 10 mW heating power. The deviation from the mean value is 2%, with larger deviations for higher temperatures. A general trend is, that the simulated membrane temperatures are lower than the measured ones. Nevertheless, for an input power of 60 mW, producing a ΔT of approximately 300 °C, the relative discrepancy between measured and simulated values is still less than 5%. As the temperature dependence of the thermal conductivities is larger at higher temperatures, there may incur additional deviations, since the temperature behavior was only determined for temperatures up to 125 °C [103], so that the heat conductivity increase may be overestimated. According to Eq. (3.26), the overall stored heat is calculated by integrating the temperature field over the geometry model regions and multiplication with the specific heat capacities summarized in Table 4.2. The temperature dependence of the

Table 4.2. Thermal conductivities and heat capacities as used for the FEM simulations

material	thermal conductivity at 300 K $\kappa_0 \, [\mathrm{Wm^{-1}K^{-1}}]$	temperature coefficient $\alpha \, [10^{-3}\mathrm{K^{-1}}]$	Ref.	specific heat capacity c $[\mathrm{MJm^{-3}K^{-1}}]$ at 300 K	Ref.
dielectric membrane	1.2	1.1	[103]	1.7 ± 0.2	[112]
polysilicon	37	0.8	[103]	1.7 ± 0.2	[112]
metal	190	1.0	[103]	2.5 ± 0.2	[112]
Si	150	-	[106]	1.7	[113]
SnO_2	28.5–30.6	-	[114]	2.6	[114]

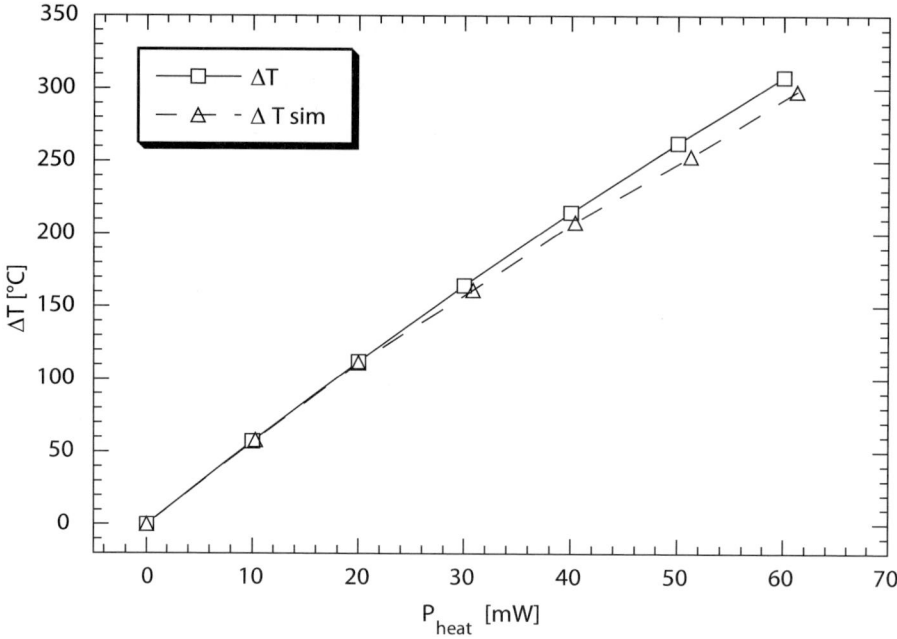

Fig. 4.5. Measurements and simulation of the temperature increase, ΔT, versus heating power of a circular microhotplate

heat capacities are neglected in the linear regime around ambient temperature. The values of the heat capacities of the different materials were originally determined for CMOS materials from a different foundry, but since all values are in agreement with tabulated bulk material values [112], they were also applied to this CMOS process. Reviewing the data, a value of $1.7 \pm 0.2 \, \text{MJ}/°\text{C} \cdot \text{m}^3$ was selected for Si and all dielectric materials. For an input power of 10 mW, which corresponds to a membrane temperature increase of $\Delta T = 57 \, °\text{C}$, the overall stored heat is calculated to be 108 µJ which corresponds to $c_0 = 1.9 \, \text{µJ}/°\text{C}$. The calculated time constant according to Eq. (3.27) is 10.8 ms.

The experimentally determined time constant was $\tau_0 = 9.7 \pm 0.2 \, \text{ms}$. The calculated time constant is approximately 11% higher than the measured one. This is a good agreement given the fact that an error of 15% was assumed as a consequence of the uncertainty of $\pm 0.2 \, \text{MJ}/°\text{C} \cdot \text{m}^3$ in the heat capacities.

In conclusion, simulated and measured values are in good agreement, and the achieved accuracy is sufficient for system-level simulations. The experimental results for the characteristic data of a circular microhotplate design are listed in Table 4.3.

The deviation between the time constants for membrane heating and cooling was measured as well (Eq. (3.37)). The heater of a single microhotplate was driven with a rectangular-shape current pulse. The pulse amplitude was adjusted to produce a temperature rise of 50 °C. In this case the measured time constant for cooling was

Table 4.3. Characteristic data of a circular microhotplate design

	circular microhotplate	circular microhotplate with sensitive layer
η_0 [°C/mW]	5.8 ± 0.2	5.8 ± 0.2
τ_0 [ms]	9.7 ± 0.2	21 ± 1

$\tau_0 = 9.7$ ms. Equation (3.37) predicts $t_{\mathrm{eff}} = 10.1$, which is in excellent agreement with the measured rise-time constant of 10.1 ± 0.1 ms.

As long as the coating is predominantly located in the membrane center, no additional heat conduction paths will be created from the heated area to the bulk silicon, so that no major changes in the thermal resistance have to be expected. The thermal time constant, however, will considerably change owing to the additional volume of the tin dioxide droplet. The time constant of a coated circular microhotplate is 21 ± 1 ms. A rough estimation for a circular droplet with a height of 25 μm and a radius of 150 μm, 80% solid material and a specific heat capacitance of the crystalline SnO_2 of 2.6 MJ/°C·m³ (Table 4.2) leads to an additional heat capacity of 1.9 μJ/°C. This corresponds to an 11-ms increase in the time constant, which is fully consistent with the measurements.

4.2 Assessment of Microhotplate Temperature Distributions

4.2.1 Device Description

The electrochemical etch-stop technology that produces the silicon island is rather complex, so that an etch stop directly on the dielectric layer would simplify the sensor fabrication (Sect. 4.1.2). The second device as presented in Fig. 4.6 was derived from the circular microhotplate design and features the same layout parameters of heaters and electrodes. It does, however, not feature any silicon island. Due to the missing heat spreader, significant temperature gradients across the heated area are to be expected. Therefore, an array of temperature sensors was integrated on the hotplate to assess the temperature distribution. The temperature sensors (nominal resistance of 1 kΩ) were placed in characteristic locations on the microhotplate, which were numbered T_1 to T_4.

The measurement results have been compared to the corresponding values of a FEM simulation in Sect. 4.2.2, and the validity of the model for simulations of the temperature distribution has been established.

Instead of a silicon island underneath the dielectric layer, a polysilicon plate can be placed in the membrane center. Such a device was not fabricated, but the effect of a heat spreader that is integrated in the dielectric membrane was demonstrated by simulations. The results of the simulations are discussed in Sect. 4.2.2 [115, 116].

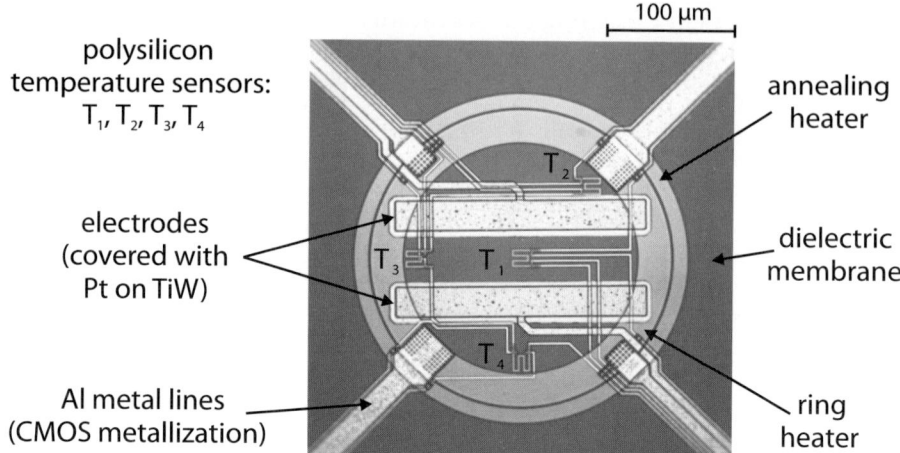

Fig. 4.6. Close-up of the circular microhotplate with temperature sensor array and without silicon island

Another issue is how the tin-oxide droplet changes the temperature distribution in the heated area. This issue was experimentally investigated and the results are summarized in Sect. 4.2.3.

4.2.2 Comparison of Simulations and Measurements

The comparison of simulation and measurement data of an uncoated membrane is shown in Fig. 4.7. The temperature curves, T_1 to T_4, were measured with the on-membrane temperature sensors. The graphs of the simulated temperatures are denoted S_1 to S_4. The temperature discrepancy between simulation and experiment was less than 5% for all sensors. The general shape of the temperature distribution was correctly modeled within measurement accuracy. It has to be noted that no additional fitting parameters were used for these simulations.

Figure 4.8 shows the relative temperature difference of the various sensors with respect to the sensor in the membrane center T_1, which acts as a reference. The values of T_2 represent, e.g., the relative difference $(T_2 - T_1)/T_1$. One would intuitively expect, that T_1 shows the lowest temperature owing to the ring heater scheme, which would lead to a positive difference value for all other sensors. However, T_2 shows a lower temperature than T_1 owing to the fact that T_2 is in close proximity to the wide metal line of the heater. As a consequence of the large heat flux through the heater line to the bulk silicon, the measured temperature of T_2 is lower. The deviation of T_4 is close to 15% and that of T_3 close to 9%.

In conclusion, the measurements support the presented microhotplate model and simulation approach. As the active region of the sensitive material is between the two electrodes, T_3 is a good measure for the temperature gradient across the active region. Although this device has no feature for improving the temperature homogeneity (silicon island etc.), the temperature gradient at T_3 at 300 °C hotplate temperature is only

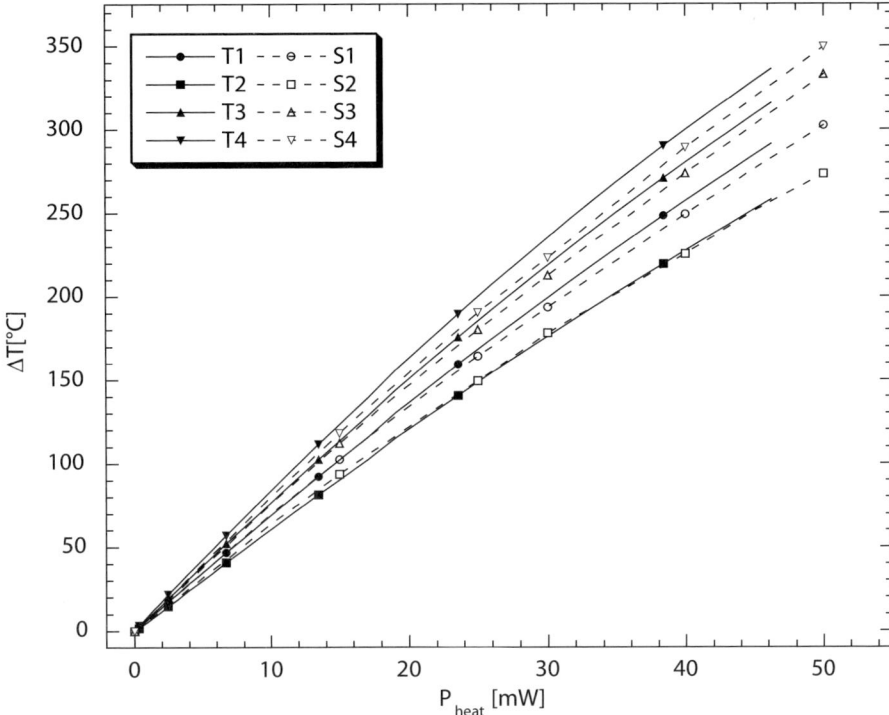

Fig. 4.7. Simulated and measured temperature distribution in the heated area of a microhotplate without Si-island. T_1 to T_4 denote the temperature sensor measurements, S_1 to S_4 the simulated temperatures at the respective locations

0.3 °C/µm. With a Si-island underneath, the temperature homogeneity is further improved. For a comparable device with a Si heat spreader a relative deviation of less than 2% equivalent to a temperature gradient of 0.07 °C/µm at 300 °C in the active area was achieved (see Sect. 4.4.4 and [81]).

Another possibility to improve the temperature homogeneity is to introduce an additional polysilicon plate in the membrane center. The thermal conductivity of polysilicon is lower than that of crystalline silicon but much higher than the thermal conductivity of the dielectric layers, so that the heat conduction across the heated area is increased. Such an additional plate constitutes a heat spreader that can be realized without the use of an electrochemical etch stop technique. Although this device was not fabricated, simulations were performed in order to quantify the possible improvement of the temperature homogeneity. The simulation results of such a microhotplate are plotted in Fig. 4.9. The abbreviations S_1 to S_4 denote the simulated temperatures at the characteristic locations of the temperature sensors. At the location T_2, the simulated relative temperature difference is 5%, which corresponds to a temperature gradient of 0.15 °C/µm at 300 °C.

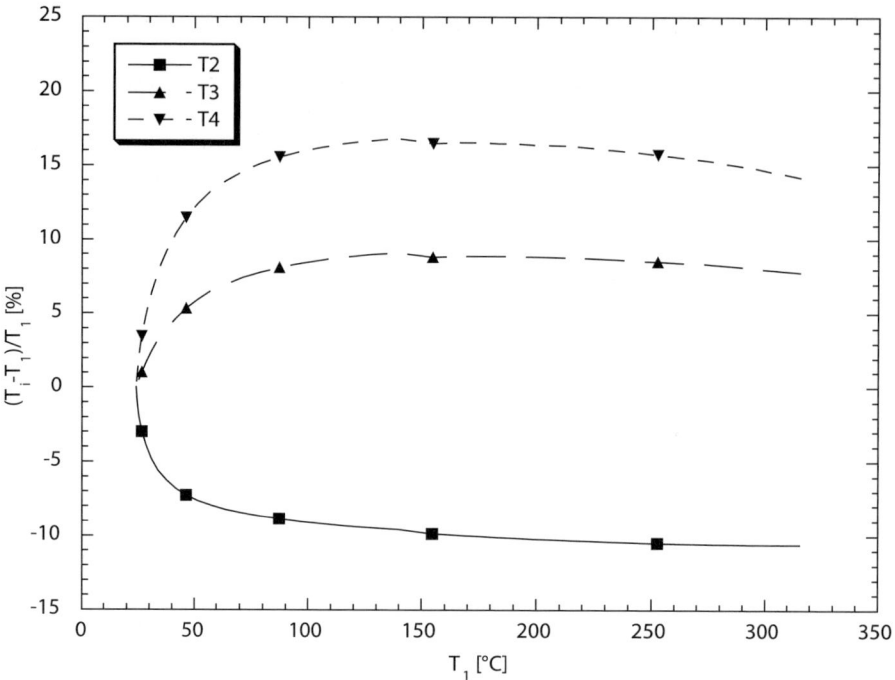

Fig. 4.8. Relative temperature differences between the temperature sensors T_2 to T_4 and T_1

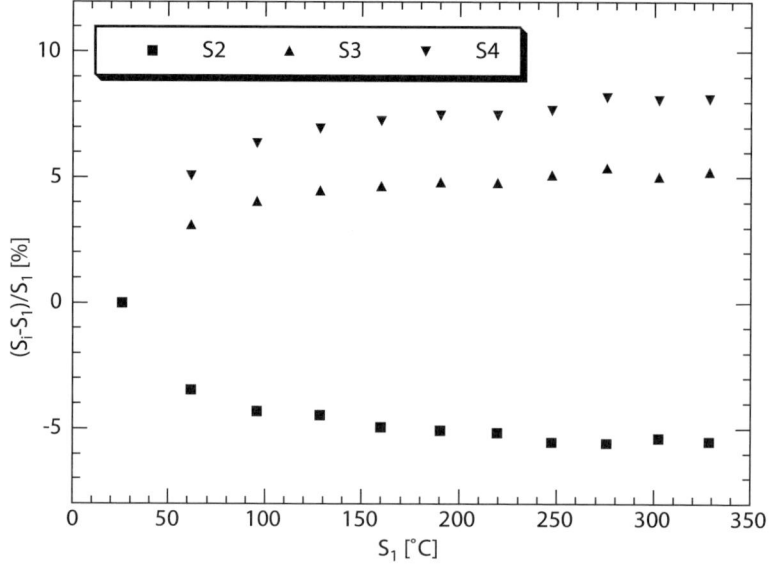

Fig. 4.9. Simulated relative temperature differences, S_2 to S_4, between the different temperature sensor locations, T_2 to T_4, and T_1. The simulated sensor features a polysilicon plate in the center

4.2.3 Temperature Distribution Assessment of a Coated Microhotplate

Since the bare uncoated membrane represents a worst-case-scenario in terms of temperature homogeneity, it is mandatory to analyze the effect of the tin-oxide droplet on the temperature distribution. The material constants of the SnO_2 thick-film layer are not well known, and the deposition is not as reproducible as the industrial CMOS process and the subsequent micromachining steps. Therefore, the sensitive layer was not included in the simulations. The heat conductivity and heat capacity of crystalline SnO_2 are listed in Table 4.2. These values might scale with the porosity of the material, but it is also known, that the heat conductivity is strongly reduced in nano-porous materials such as porous silicon [117]. Since the SnO_2 droplet has a typical height of 25 μm, the temperature homogeneity will be increased by heat spreading through the thick layer. The active area of the material is almost identical with the heated part of the microhotplate surface, so that the simulations of the non-coated microhotplate can be taken as a coarse approach to finding the temperature distribution on the coated microhotplate.

The temperature distribution on a coated microhotplate was experimentally assessed. The results of the measurements are plotted in Fig. 4.10. In comparison to the results presented in Fig. 4.8, the temperature gradients at all temperature sensor locations were lower as can be expected. An interesting feature was the change in sign for temperature sensor T_2. An interpretation of this change is, that the heat flux via the SnO_2 droplet provides enough heat to counterbalance the heat loss through the metal line, so that the temperature difference between T_2 and T_1 is reduced. The relative discrepancy between temperature sensors T_3 and T_1 amounts to half of the discrepancy for the uncoated device and is close to 3% at 300 °C.

The conclusion from the results of this chapter is, that a silicon island fabricated by ECE is not absolutely necessary, if a relative temperature difference of 5% within the active area between the electrodes is acceptable. A microhotplate with a dielectric membrane and a polysilicon heat spreader in the center features sufficient temperature homogeneity. Moreover, the tin-oxide droplet serves as additional heat spreader and smoothes out temperature gradients.

In conclusion, a simple KOH-etching process without ECE is applicable for future microhotplate designs, although the best temperature homogeneity is achieved with the silicon island heat spreader. The island remains an important design feature, especially for the use of thin-film sensitive layers, where the additional heat spreading effect of the sensor materials is small.

4.3 Microhotplate with Pt Temperature Sensor

4.3.1 Design Considerations

The circular microhotplate presented in Sect. 4.1 features an upper sensor operating temperature limit of 350 °C, which is imposed by the CMOS metallization. At higher temperatures, electromigration, especially in the heater structures, will occur,

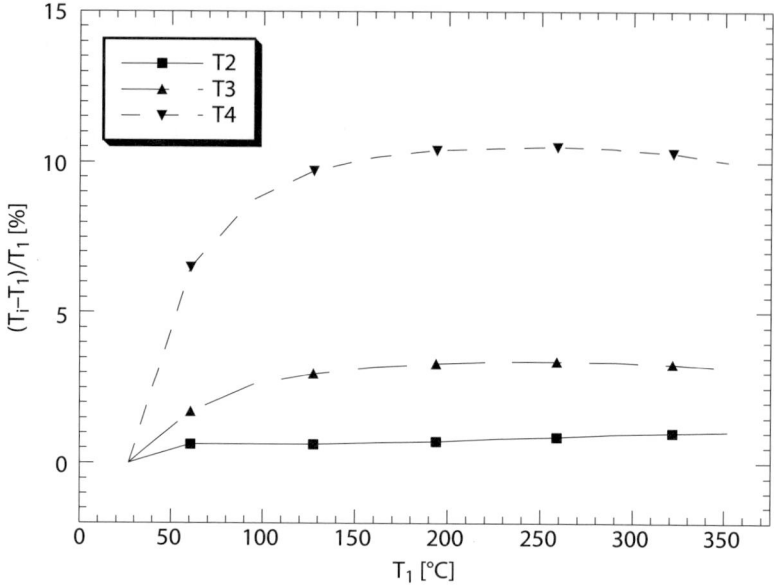

Fig. 4.10. Measured relative temperature differences between the temperature sensors T_2 to T_4 and T_1. The microhotplate was coated with a nano-crystalline SnO_2 droplet

which limits the life time of the system. Thus, any contacts and metal lines in the heated area have to be avoided. Another problematic element is the poly-Si temperature sensor.

It is known, that polycrystalline Si changes its resistance at higher temperatures. The changes in the resistance are either attributed to a recrystallization process [118] or degradation due to interlayer diffusion and thermomechanical stress defects [119]. Such a resistance change will lead to a drift in the temperature sensor signal. A resistance drift will be also observable in a polysilicon heating resistor, it is, however, less critical in comparison to that of the temperature sensor, since the temperature controller adjusts the heating current according to the temperature feedback and automatically compensates for changes in the heating resistor.

Another issue is the membrane buckling due to internal and thermal stress of the CMOS layers, which are not optimized for high-temperature operation. The buckling of the microhotplate can generate severe problems in the adhesion of the sensitive layer. In a temperature-pulsed operation mode, the repeated bending of the membrane could cause a reliability problem [120].

A novel microhotplate design was proposed to overcome the CMOS operating temperature limit and to avoid polysilicon-induced drift problems. A cross-sectional schematic of the device is shown in Fig. 4.11. Instead of using a polysilicon resistor as temperature sensor, a platinum temperature sensor is patterned on the microhotplate. The Pt-metallization process step was used to simultaneously fabricate the electrodes and the temperature sensor. The CMOS-Al/Pt contacts are located off the membrane

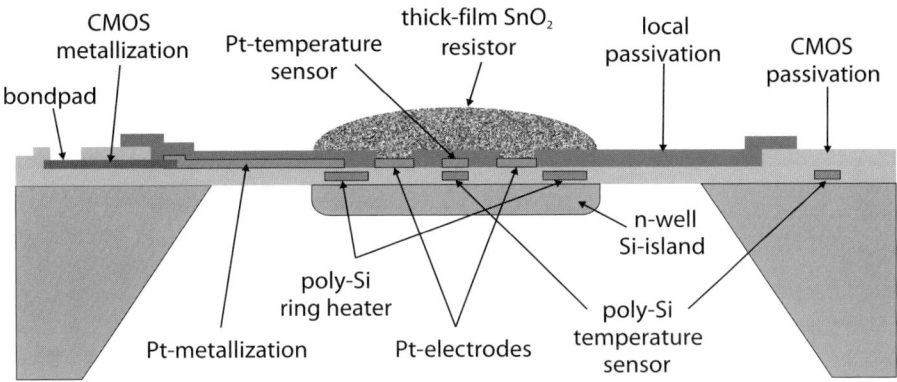

Fig. 4.11. Cross-sectional schematic of the high-temperature microhotplate

on the bulk chip and no more in the heated membrane center. The Pt-elements have to be protected by a passivation layer, and the passivation has to be opened again to contact the sensitive layer to enable resistive measurements. The passivation layer is deposited locally, which allows to use materials that reduce or compensate the mechanical stress of the membrane. A stress compensation technique has been used on wafer level to reduce the overall stress in the dielectric layers with the drawback, that the stress characteristics of the passivation have to be modified within the CMOS process [121, 122]. The CMOS passivation layer, however, is primarily optimized for the protection of the circuitry. Applying local passivation depositon prevents, that CMOS transistor parameters change upon passivation modifications, or that the device reliability is compromised. Moreover, the thickness of the local passivation can be different from that of the CMOS passivation, and the deposition process can be developed independently of the CMOS foundry.

Figure 4.12 shows the high-temperature microhotplate design, and Table 4.4 summarizes the design parameters. The Pt-resistor was connected in a 4-point configuration with the Pt-Al-contacts located on the bulk silicon. An additional poly-Si temperature sensor was integrated underneath the Pt-resistor, which can be used as a reference. The heater is still fabricated in polysilicon, but the design was modified in order to avoid Al-contacts in the heated area. The heater has a nominal resistance of 200 Ω and was, again, realized in a parallel configuration. The contact of the polysilicon to the metallization is not located in the heated area, but at a certain distance from the heated area on the membrane. This contact position features a reduced contact temperature as is illustrated in Fig. 4.13.

On the right-hand side of Fig. 4.13 the temperature distribution along two paths across the membrane is schematically plotted. The first path is along a metal line. Since only heat conduction along the metal line occurs, a linear temperature decrease is observed. The second path is along the heater connection. Owing to Joule-heating and the lower thermal conductivity of the resistive poly-Si section of the connecting arm, a significant temperature drop occurs. The temperature of the contact is considerably lower in comparison to path 1 at the same distance from the hotplate center.

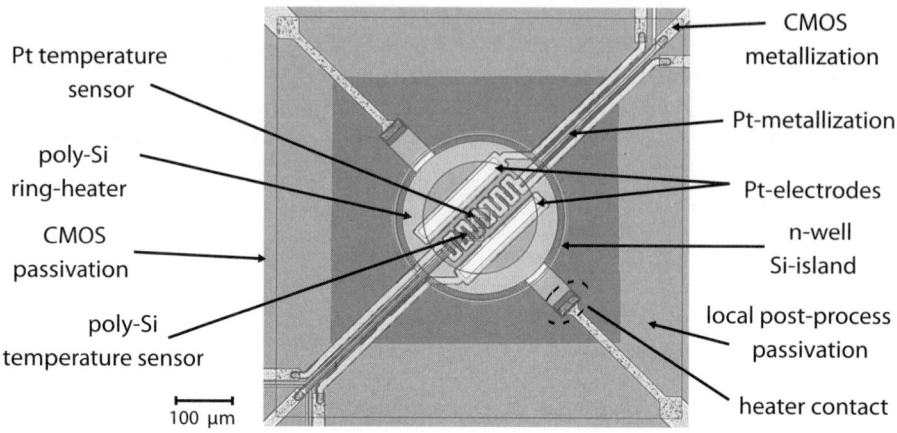

Fig. 4.12. Micrograph of the microhotplate with Pt temperature sensor

Table 4.4. Design parameters of the high-temperature microhotplate

membrane size	$500 \times 500\ \mu m^2$
Si-island diameter	$300\ \mu m$
heater resistance	$200\ \Omega$
temperature sensor resistance	$75\ k\Omega$
reference resistor	$10\ k\Omega$
electrode distance	$80\ \mu m$
electrode length	$185\ \mu m$

As a consequence, the temperature slope in the metal-line part of path 2 between the contact and the bulk chip is lower, which indicates a reduced heat flow through the metal line. This is equivalent to a better thermal decoupling of the metal power supply lines from the heated area, so that the presented microhotplate design is well suited to achieve operating temperatures up to 500 °C.

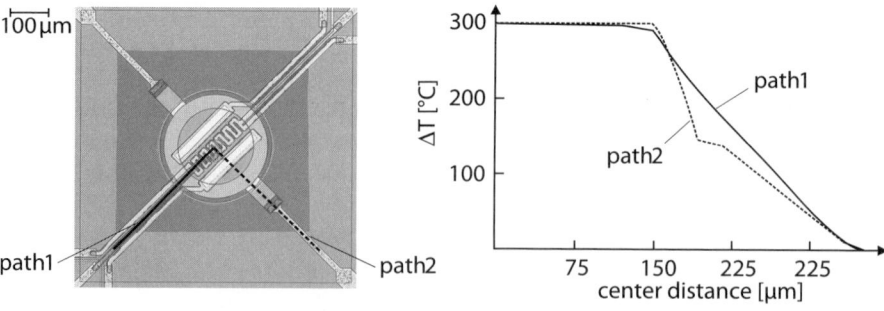

Fig. 4.13. Temperature distribution along a metal connection path (path 1) and a heater contact arm (path 2)

4.3.2 Fabrication

A process sequence schematic of the principal steps is given in Fig. 4.14. The process sequence is similar and compatible to the fabrication of the circular hotplate in Sect. 4.1.2. The most important modifications include the additional process steps to apply the local passivation. The modifications of the other steps are described in the following section [115].

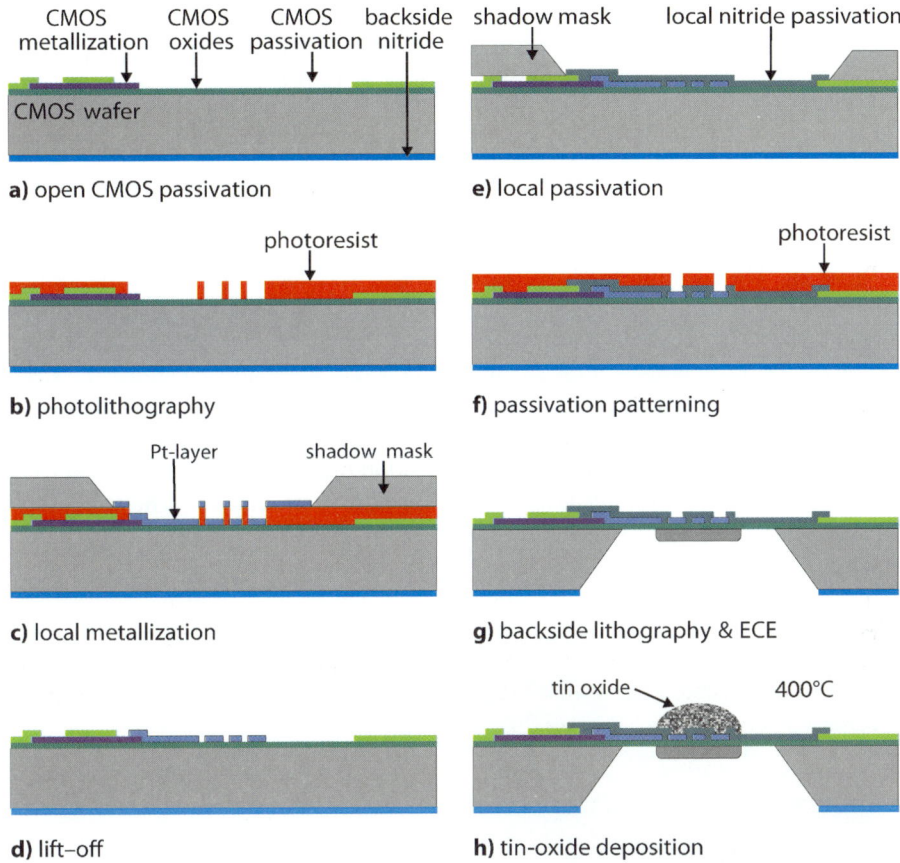

Fig. 4.14. Schematic of the microfabrication process for the high-temperature microhotplate with Pt temperature sensor

Local Metallization

The standard CMOS passivation is locally opened already during the CMOS process (Fig. 4.14a). Open CMOS metallization structures serve as a contact to the Pt-wires. The thickness of the TiW-layer can be reduced to 20 nm, since there is no aluminum underneath, and the metal stack is directly sputtered onto the PECVD intermetal oxide of the CMOS process (Fig. 4.14c). The applied lift-off process (Fig. 4.14d) is identical with the procedure described in Sect. 4.1.2.

Local Passivation

After the lift-off process, the surface of the wafer is exposed to an O_2-plasma in a plasma asher to remove possible resist residues, and to activate the surface for the subsequent nitride deposition.

To assure local deposition, the PECVD silicon nitride is deposited through a shadow mask. After alignment, the CMOS wafer and the shadow mask are fixed by a wafer holder to assure a close contact between mask and substrate. This has the advantage, that nitride deposition underneath the edges of the shadow mask is marginal, and that the boundaries of the deposited nitride area are well defined. In Figure 4.15 a micrograph of a nitride layer deposited on a bare silicon wafer is shown. As can be seen from the interference pattern, the thickness in the membrane center is fairly homogeneous. The layer thickness decreases towards the edges. Little nitride deposition is observed along the frame edges within a distance of 75 μm for a 1.5 μm thick silicon nitride layer.

Mixed-frequency deposition of the nitride is one possibility to adjust the stress in the deposited layer [122]. The ratio of the deposition times in the high-frequency (375 kHz) and low-frequency (187.5 kHz) plasma can be varied during the process. For the layer used here 95% high-frequency deposition time was chosen (Fig. 4.14e).The stress was measured on wafer-level with a thin-film stress analyzer. The stress value was determined by recording the curvature of the wafer after thin-film deposition. A tensile stress of 75 ± 5 MPa was measured for the layer.

250 μm

Fig. 4.15. Silicon nitride spot locally deposited through a shadow mask on a bare Si-substrate

After the deposition of the passivation, the contact electrodes for the resistive measurements have to be opened (Fig. 4.14f). The respective nitride etching was done in a reactive-ion etcher under SF_6 atmosphere.

The membrane release and the deposition of the sensitive layer are identical with the process described in Sect. 4.1.2.

4.3.3 Thermal Characterization

The fabricated microhotplate chips were packaged in a DIL-28 package and bonded. Calibration and thermal characterization of the Pt-temperature sensor was carried out according to the procedures described in Sect. 4.1.3 and 4.1.4. The thermal resistance and thermal time constant values averaged over three identical uncoated microhotplates are summarized in Table 4.5. The thermal resistance is 30% larger as compared to the circular membrane. The increase is a consequence of reduced Al-metallization line width, as the high-temperature microhotplate does not exhibit any annealing heater. As discussed in one of the previous sections, the heater design also entails a reduced heat flow through the power supply lines. A higher thermal resistance is desirable, since the higher operating temperatures can be achieved at lower overall power consumption.

Table 4.5. Characteristic thermal data of a microhotplate with Pt temperature sensor

	Microhotplate with Pt temperature sensor
η_0 [°C/mW]	7.6 ± 0.2
τ_0 [ms]	10.2 ± 0.2

4.4 Microhotplate with MOS-Transistor Heater

4.4.1 Basic Heater System Architectures

For the monolithic integration of microhotplates with on-chip temperature controllers two basic approaches are possible. Most microhotplates rely on a resistor as heating element as discussed in the previous chapter and as is shown in Fig. 4.16a. Driving the heating current according to the controller output requires an on-chip power transistor. However, the voltage-drop across this transistor leads to a massive fraction of the overall power being dissipated in the power transistor, which is not contributing to heating the hotplate. Alternatively, CMOS technology allows to directly utilize a transistor as heating element so that all power contributes to heating the hotplate (Fig. 4.16b). The power consumption of the microhotplate can be reduced in this approach, and novel heating schemes become feasible.

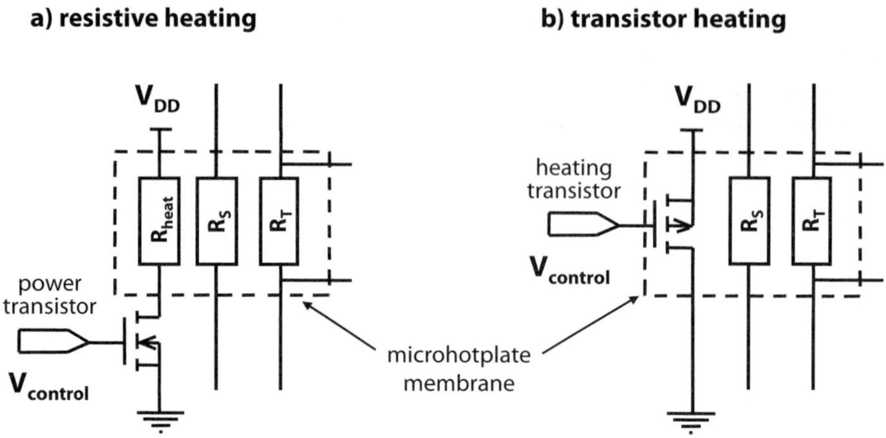

a) resistive heating   **b) transistor heating**

Fig. 4.16. Heating approaches for monolithically integrated microhotplates (μHP): (a) shows a resistive heater with power transistor and (b) shows a PMOS transistor heater: R_{heat} denotes the heating resistor; R_S is the metal-oxide chemiresistor, and R_T is a resistor used as temperature-sensor (see Fig. 4.4)

Microhotplates with an active transistor heating element have been fabricated in SOI (Silicon on insulator)-CMOS technology [22, 52]. SOI technology is commonly used for devices that have to work reliably at operating temperatures higher than 120 °C. No examples of microhotplates in conventional CMOS technology have been reported on so far, which is probably due to the fact that an increased leakage current occurs in the diode insulation to the bulk material at elevated temperatures [123, 124]. To circumvent this problem, it is possible to fabricate a microhotplate with a n-well Si-island in the center using the electrochemical etch stop technology, that has been presented in Sect. 4.1. A PMOS heating transistor can be placed on the n-well island, which is thermally and electrically insulated from the bulk, so that leakage currents to the bulk material are suppressed, and the transistor heating element preserves its basic transistor properties even at higher temperatures. An analytic model based on standard transistor equations has been introduced to describe the properties of such a microhotplate. A main goal was to only use device parameters, which are available from the CMOS foundry. The use of additional fitting parameters was avoided, since these would compromise the compatibility with standard circuitry modelling. The identification of important parameters and effects that influence the thermal characteristics of the microhotplate was the primary goal. The model with some simplifications led to a sufficiently accurate description of the device.

4.4.2 System Description

A schematic view of the microhotplate with transistor heater is shown in Fig. 4.17 [125]. In order to ensure a good thermal insulation, only the dielectric layers of the CMOS process form the membrane. The inner section of the membrane includes an n-well silicon island underneath the dielectric layers. The n-well is electrically insulated and serves as heat spreader due to the good thermal conductivity of silicon. It also hosts the PMOS-transistor heating element. The n-well source and drain contact diffusion are indicated in the cross-sectional drawing (Fig. 4.17).

A special octagonal-shape transistor arrangement was chosen for homogeneous heating (Fig. 4.18). A polysilicon resistor is used to measure the temperature on the microhotplate. The resistance of the SnO_2 thick-film layer is measured using two metal electrodes. The inner section of the membrane ($500 \times 500 \, \mu m^2$) exhibits an octagonal-shape n-well silicon island ($300 \, \mu m$ base length). The octagonal shape provides a comparatively long distance between the heated membrane area and the cold

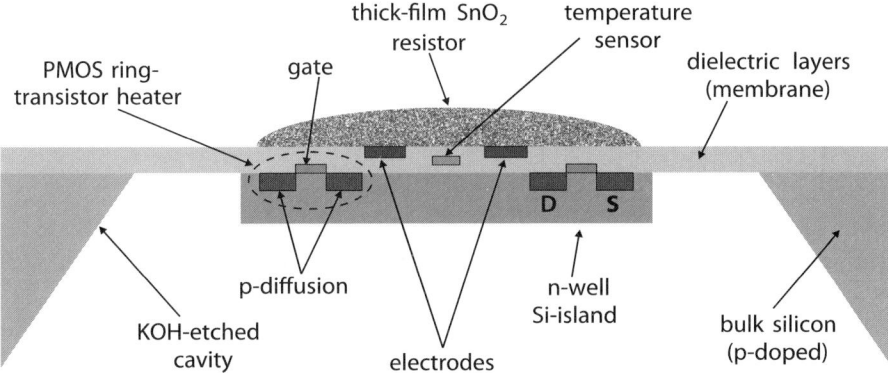

Fig. 4.17. Cross-section of a microhotplate with integrated MOS-transistor heater, D and S denote drain and source of the transistor

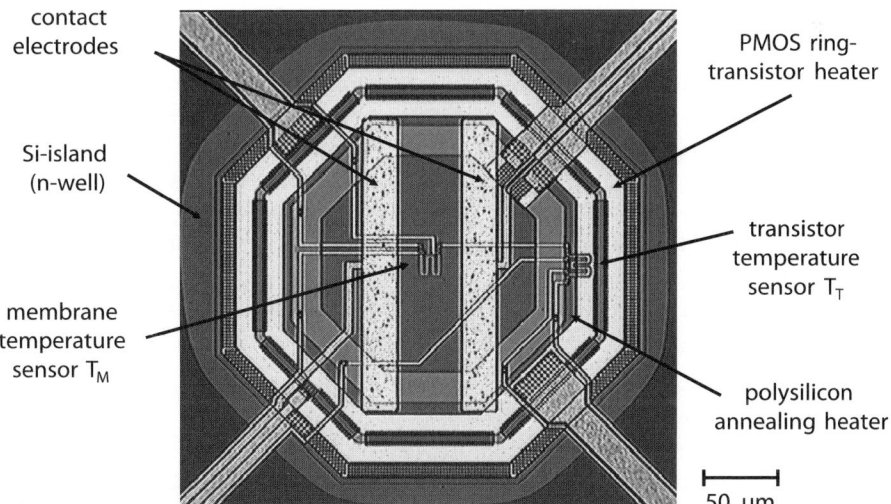

Fig. 4.18. Micrograph showing the membrane and the PMOS-heater

bulk chip, which considerably reduces the heat dissipation via metal connections in comparison to a quadratic layout of the island (see Fig. 4.18).

A PMOS-ring transistor of 5 μm gate length and 710 μm overall gate width is integrated in the n-well as the active heating element. As the major heat losses occur along the edges of the silicon island, a ring-like transistor geometry promotes homogeneous heat distribution. The ring-like heater structure also offers the advantage, that it leaves some space for additional functional elements. In addition to the PMOS transistor, two resistive temperature sensors are implemented in polysilicon, one in the center and the other near the transistor (Fig. 4.18). By measuring the membrane temperature at two locations, the heat distribution on the membrane can also be roughly assessed. An additional resistive heater is implemented on the membrane as well. This heater was designed for in-situ annealing of the sensitive material. As the sensitive layers used in this work are annealed in a furnace as described in Sect. 4.1.2, the extra annealing heater was not used. The fabrication of the chip is identical to that of the circular microhotplate described in Sect. 4.1.2.

4.4.3 Analytical Model for the MOS-Transistor Heater

Starting with the basic model assumptions, the analytical heater model is developed in several steps [126]. The equations include common model equations such as the Shichman-Hodge model [127], the LEVEL3 model [128] and the BSIM3.v3 model [129]. Only selected components of these partly complex models were taken to yield a set of equations that is suitable for modelling the transistor heater. The variables and parameters have been defined in accordance to standard notations. First, a model has been established that describes the unheated transistor, then, temperature dependencies have been introduced, and, finally, the electrothermal coupling to the microhotplate has been considered. The result is an implicit equation, which can be iteratively solved. The considered model will be compared to measurement data in Sect. 4.4.4.

To achieve membrane temperatures higher than 300 °C, a thermal power of several tens of mW is required, which entails a source-drain current of several mA at a source-drain voltage of 5 V. For the transistor used here featuring a channel length of 5 μm and an overall gate width of 710 μm, it is legitimate to largely neglect short-channel and narrow-channel effects. The transistor works in strong inversion and is exclusively driven in the saturation region so that the basic transistor equation holds:

$$I_{sd,sat} = \frac{\mu_{eff} \cdot W_{eff} \cdot C_{ox}}{2 \cdot L_{eff}} \cdot \left(V_{sg} - V_t\right)^2 \cdot \left(1 + \lambda \cdot V_{sd}\right) \qquad (4.3)$$

The saturation current is denoted $I_{sd,sat}$, the effective mobility μ_{eff}, and the capacitance of the gate oxide C_{ox}. The effective geometrical parameters of the transistor include its overall gate width, W_{eff}, and the effective gate length, L_{eff}.

To be consistent with common equations for NMOS transistors, the polarities of the applied voltages have been inverted. The source-gate voltage is denoted V_{sg}, the source-drain voltage V_{sd}, and the threshold voltage V_t. Equation (4.3) corresponds to the analytical Shichman-Hodges MOSFET model, the first two terms of which represent the "square-law" description of the saturated drain current [127]. The last term is a simple expression for the effect of channel-length modulation, where λ is extracted from measured data. The effective mobility, μ_{eff}, takes into account the modifications of the low-field mobility, μ_0, owing to the electrical fields in the transistor. The mobility is, therefore, subdivided into three components:

- The low-field mobility μ_0
- The mobility, μ_v, which represents the modifications of μ_0 due to the vertical electric field;
- The mobility, μ_{eff}, which represents the modifications of μ_v due to the lateral (channel) electric field.

The effects of vertical and lateral fields are assumed to be independent of each other. The modification of the mobility due to vertical field is expressed as [BSIM3.v3, 129]:

$$\mu_v = \frac{\mu_0}{1 + (UA + UC \cdot V_{\text{sbx}}) \cdot \left(\dfrac{V_{\text{sg}} + 2 \cdot V_{\text{t}}}{t_{\text{ox}}}\right) + UB \cdot \left(\dfrac{V_{\text{sg}} + 2 \cdot V_{\text{t}}}{t_{\text{ox}}}\right)^2} \tag{4.4}$$

In this equation UA, UB, and UC are a set of parameters from the foundry for the respective CMOS technology. V_{sbx} denotes the potential difference between source and bulk, which is zero in the case of the MOSFET-heater. Finally t_{ox} represents the thickness of the gate oxide.

The effective mobility can then be written as [LEVEL3, 128]:

$$\mu_{\text{eff}} = \frac{\mu_v}{1 + \mu_v \cdot V'_{\text{sd}} / (v_{\text{sat}} \cdot L_{\text{eff}})} \tag{4.5}$$

where $V'_{\text{sd}} = \min(V_{\text{sd}}, V_{\text{sd,sat}})$ and $V_{\text{sd,sat}} = V_{\text{sg}} - V_{\text{t}}$ is the source-drain saturation voltage, the value of V_{sd}, at which the channel pinches off. Since the transistor is exclusively driven in the saturation region, the relation $V'_{\text{sd}} = V_{\text{sg}} - V_{\text{t}}$ holds. The saturation velocity is given by v_{sat}.

These equations describe an unheated transistor and were verified for a device with no backside etching (no membrane). The modelling parameters were provided by the manufacturer, whereas the value of the threshold voltage was taken from wafer map data. The channel length modulation parameter, λ, had to be extracted from measurement data. The discrepancy between simulated and measured source-drain saturation current, $I_{\text{sd,sat}}$, for a transistor embedded in the bulk silicon was less than 1%, which confirmed the validity of the model assumptions.

The next step was the introduction of the temperature dependence of the relevant parameters. A linear approximation was chosen for the temperature dependence of the threshold voltage:

$$V_{\text{t}}(T) = V_{\text{t}}(T_{\text{nom}}) - TCV \cdot (T - T_{\text{nom}}) \tag{4.6}$$

where TCV is the temperature coefficient of the threshold voltage and T_{nom} represents the reference temperature, at which the parameters are extracted. For the device under investigation T_{nom} is 25 °C. Equation (4.6) reflects the fact, that the threshold voltage decreases with increasing temperature.

The temperature-dependence of the low-field mobility can be described as:

$$\mu_0(T) = \mu_0(T_{nom}) \cdot \left(\frac{T}{T_{nom}} \right)^{-1.5} \tag{4.7}$$

The temperature sensitivity of the saturation velocity, v_{sat}, used in Eq. (4.5) to describe the effect of the lateral field, is additionally considered. The saturation velocity is moderately affected by temperature changes; however, when the temperature largely deviates from T_{nom}, there is a strong effect on the saturation region characteristics of the transistor. The temperature dependence of the saturation velocity is taken into account using the temperature coefficient, AT, [BSIM 3.v3, 129]:

$$v_{sat}(T) = v_{sat}(T_{nom}) - AT \cdot \left(\frac{T}{T_{nom}} - 1 \right) \tag{4.8}$$

Replacing the respective variables in Eq. (4.3) using the Eqs. (4.5), (4.6), (4.7) and (4.8), a temperature-dependent MOS transistor model is obtained. This temperature-dependent model provides a term for the source-drain current depending on the source-gate voltage, the source-drain voltage, and the temperature.

The last step in the construction of the MOSFET-heater model includes the description of an appropriate heating process. Due to the source-drain current flow, the membrane is heated by resistive Joule heating in the channel region. By assuming that all electric power dissipated in the device is converted into heat, the corresponding heating power is:

$$P_{heat} = I_{sd} \cdot V_{sd} \tag{4.9}$$

Combining Eq. (4.9) with the lumped-model Eq. (3.24), one gets an expression for the temperature on the membrane depending on the source drain-current:

$$T = T_0 + \eta_0 \cdot \left(I_{sd}\left(T, V_{sg}\right) \cdot V_{sd} \right) + \eta_1 \cdot \left(I_{sd}\left(T, V_{sg}\right) \cdot V_{sd} \right)^2 \tag{4.10}$$

where T_0 is the membrane temperature for zero source-drain bias, which corresponds to ambient temperature. The remaining variables in Eq. (4.10) are the temperature of the transistor, T, as a function of the source-gate voltage, V_{sg}, at constant source-drain supply voltage. It is not possible to analytically solve this equation for T, owing to the implicit temperature dependence of I_{sd}. Therefore an iterative method has been chosen to find a solution for the temperature as a function of the source-gate voltage, V_{sg}, which has been implemented in MATHEMATICA™.

The coefficients of thermal resistance can either be measured for existing devices or be calculated with the thermal microhotplate model presented in Chap. 3. In analogy to resistor-heated membranes, the model can be used for evaluation and optimization of new designs. A combination of the presented transistor model equations with the lumped microhotplate model in Sect. 3.4 would allow to also derive an AHDL model for coupled-system simulations.

4.4.4 Electrothermal Characterization and Comparison with Simulations

The microhotplate with the transistor heater was electrothermally characterized similarly to the procedures presented in Sect. 4.1.3. Special care was taken to exclude wiring series resistances by integration of on-chip pads that allow for accurate determination of V_{sg} and V_{sd}. With the two on-chip temperature sensors in the center (T_M) and close to the transistor (T_T) the temperature homogeneity across the heated area was assessed as well. Both sensors were calibrated prior to thermal characterization. The relative temperature difference $(T_T - T_M)/T_M$ was taken as a measure for the temperature homogeneity of the membrane. The measured thermal characteristics of a coated and an uncoated membrane are summarized in Table 4.6. The experimental values have been used for simulations according to Eq. (4.10).

The majority of the transistor parameters of the microhotplate were taken directly from the map or non-public simulation data provided by the CMOS foundry austria*microsystems*. The only parameter that was experimentally determined for an unetched transistor was the channel-length modulation parameter λ. As mentioned before in the introduction to this chapter, no fitting parameters were used for the simulation. The first graph in Fig. 4.19 shows measured and simulated data for temperature versus source-gate voltage for a V_{sd} of 5 V. As can been seen in Fig. 4.19, temperatures even higher than 350 °C can be achieved with a MOS-transistor-heated membrane. The relationship between membrane temperature and source-gate voltage is almost linear. This feature is very interesting for the integration of the microhotplate in a monolithic sensor system as will be discussed in Sect. 6.3. The discrepancy between simulated values and measurements is less than 5% up to a temperature of 300 °C. The discrepancy becomes more important above 300 °C, where the limits of the model are obviously reached.

The saturation current, I_{sd}, is plotted versus the temperature in Fig. 4.20. In this case, the temperature is the independent variable, which is adjusted by the source-gate voltage. The plot thus shows how well the transistor model in Eq. (4.3) takes into account the temperature dependence.

Table 4.6. Experimentally determined values of the thermal resistance, the thermal time constant, and the temperature homogeneity for uncoated and coated, transistor-heated microhotplates

	microhotplate without sensitive layer	microhotplate with sensitive layer
thermal resistance η_0 [°C/mW]	5.8 ± 0.2	6.2 ± 0.2
thermal time constant τ_0 [ms]	9.0 ± 0.2	23.3 ± 0.2
temperature homogeneity $(T_T - T_M)/T_M$	$\approx 2\%$	$\approx 2\%$

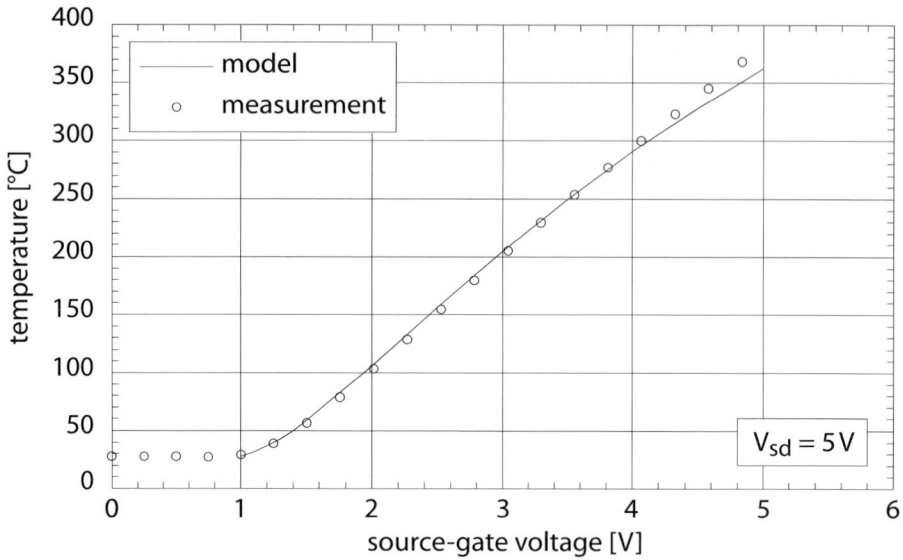

Fig. 4.19. Comparison between measured T-versus-V_{sg} characteristics and the MOSFET-heater model data for a source-drain voltage of 5 V

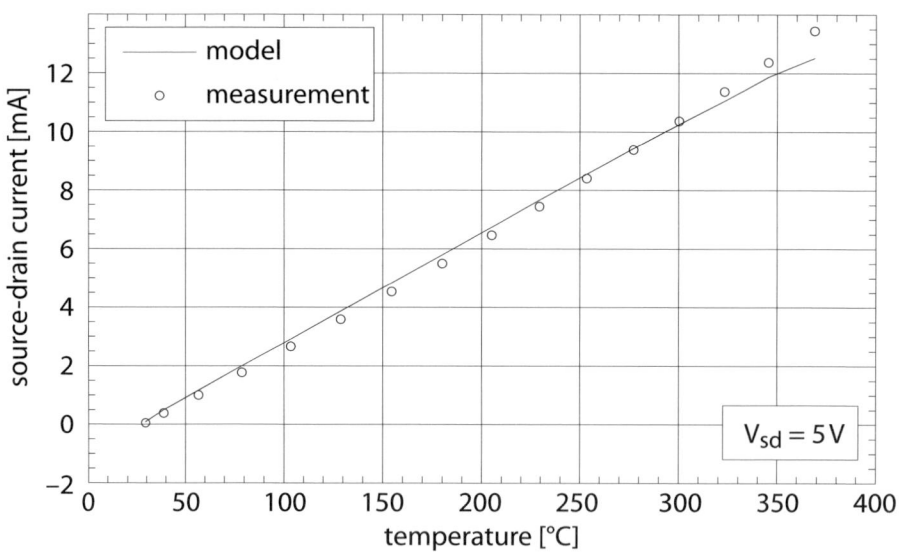

Fig. 4.20. Comparison between measured I_{sd}-versus-T characteristics of the MOS transistor and the MOS-transistor model results for a source-drain voltage of 5 V

The relative deviation between measurement results and the temperature-dependent MOS transistor model data was less than 10% above 100 °C. In the case of a source-drain bias of 5 V it appeared that the model described the real situation well up to 300 °C, but then started to deviate.

Since the uncertainty in model parameters limits the model accuracy, a deviation of 10% is generally considered a very good result. The measurements therefore demonstrate the validity of the presented model, in particular when considering that the temperature-dependent parameters are extracted from data that are only valid in a temperature range between 0 °C and 100 °C.

A second reason for the model/measurement discrepancies are the strongly temperature-dependent leakage currents of the reverse-biased p-n junction in the drain contact of the transistor heater. At room temperature, these leakage currents can be neglected, but due to their exponential temperature characteristics, they become significant when the temperature increases. As a rule of thumb, the leakage currents double approximately every 10 °C. In the source-drain current measurements done so far, the leakage current of the reverse-biased drain/n-well-junction is inherently included in the value of the source-drain current, as the leakage current also contributes to the heating process. Since the MOSFET-heater model entirely disregards the influence of leakage currents, this may lead to discrepancies between measurement data and model. To quantify this discrepancy, the leakage current at the n-well terminal has been measured at different source-gate bias voltages and has been compared with the overall current at the drain terminal (Fig. 4.21).

It is noteworthy, that, in the case of the MOSFET heater, the transistor temperature is a function of the drain current, and, therefore, the leakage current also strongly depends on the drain current. As it is shown in Fig. 4.21 the leakage current amounts

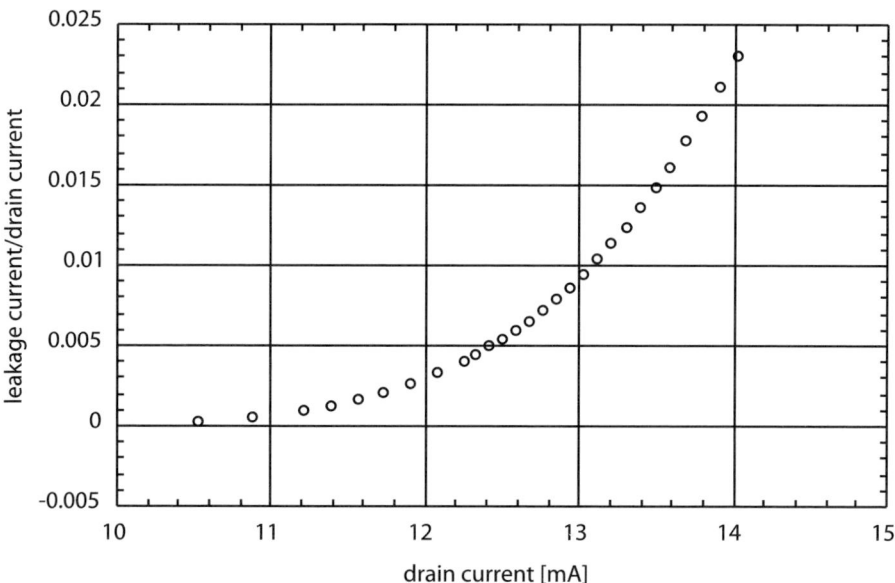

Fig. 4.21. Ratio of the leakage current of the reverse-biased drain/n-well junction and the drain current as a function of the drain current. A drain current of 14 mA corresponds to a membrane temperature of 375 °C

to almost 2.5% of the drain current at 14 mA. This current produces a temperature of 375 °C. Ignoring the leakage current during the modeling process consequently causes a temperature error in the MOS-heater model.

4.5 Calorimetric Sensing Mode for Operation at Constant Temperature

The microhotplate was coated with a thick-film tin-oxide droplet as described in Sect. 4.1.2. To characterize the chemical-sensor performance, the chip was exposed to CO concentrations from 5 to 50 ppm in humidified air at 40% relative humidity (23.4 °C water vaporization temperature) (see Sect. 5.1.8 for a description of the gas test measurement setup).

An exemplary sensor resistance plot for varying gas concentrations is shown in Fig. 4.22. As discussed in Chap. 2 detailing the chemical reactions, metal-oxide sensors also can be used as calorimetric sensors or pellistors, which monitor the temperature change during gas exposure. To improve the gas discrimination, both, the resistive signal and the temperature changes can be simultaneously monitored and evaluated. The temperature changes or calorimetric signals provide information on the thermal budget of the reactions taking place. The thermal budget and the corresponding hotplate power consumption depend on the analyte nature, the sensing material, and electrode layout. It provides additional information for better analyte discrimination. The calorimetric mode is not easy to realize with temperature-controlled microhotplates, since the purpose of the temperature controller is to keep the temper-

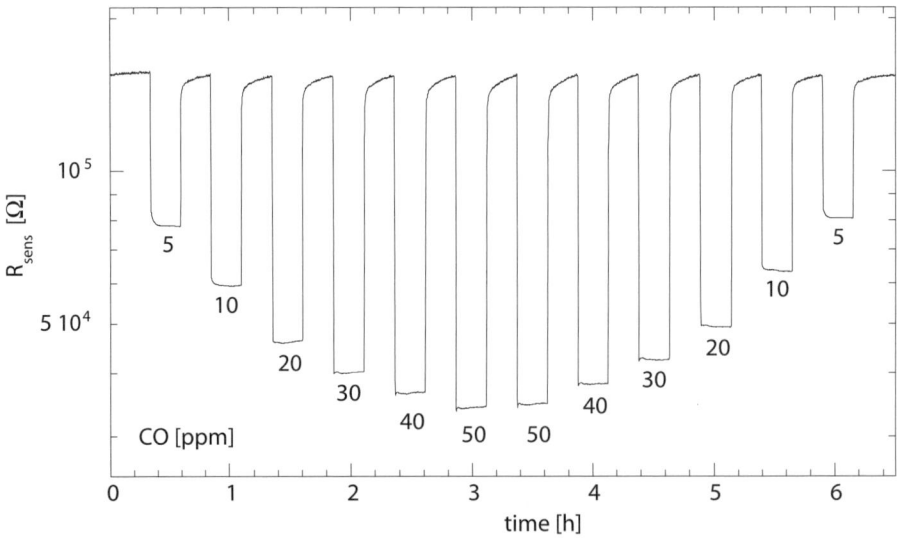

Fig. 4.22. Sensor resistance changes of a MOSFET-heated microhotplate upon exposure to CO concentrations of 5 to 50 ppm at 40% r.h.

ature constant at a preset value. The calorimetric signal is, therefore, only indirectly accessible by monitoring the heating current that is required to maintain this temperature. The MOSFET heater offers the possibility to measure the change in the source-gate voltage that is needed to maintain a preset constant temperature during analyte exposure of the sensor. The temperature control was implemented by using a Labview™ program, that reads out the resistance of the temperature sensor and readjusts the source-gate voltage accordingly so that the temperature is kept constant.

Measurement results are plotted in Fig. 4.23. The sensor signal is recorded under the same experimental conditions and with the same chip as the measurements displayed in Fig. 4.22. The membrane temperature was preset to be 300 °C. An increasing source-gate voltage requires a larger heating power to maintain constant temperature. Without any change in the heating power, the membrane temperature would decrease. A concentration step of 50 ppm results in a voltage step of 3 mV. This corresponds to a temperature change of 0.3 °C. The noise of the signal is equivalent to a temperature fluctuation of 0.02 °C.

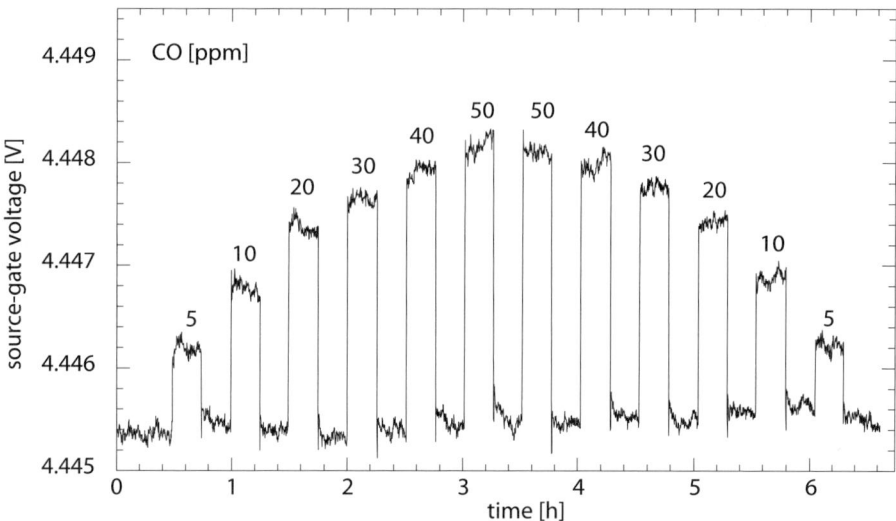

Fig. 4.23. Source-gate voltage changes of a MOSFET-heated microhotplate upon exposure to CO concentrations of 5 to 50 ppm at 40% r.h.

A control experiment was performed with an uncoated microhotplate in the gas test measurement setup with the same measurement program and at the identical microhotplate temperature. No changes in the source-gate voltage could be measured. Therefore, the changes in the microhotplate heat budget are clearly related to the interaction of the analyte with the tin oxide.

A problem of the calorimetric sensing mode is its cross-sensitivity to changes in ambient temperature. The realization of an additional temperature sensor on the bulk chip solves this problem. The signal-to-noise ratio of the calorimetric mode is less favorable than that of the chemoresistive measurements. Nevertheless, valuable calorimetric information can be gained with temperature-controlled microhotplates. However, it is difficult to implement on-chip temperature controllers that provide a temperature stability of better than $\pm 1\,^\circ$C as will be shown in the following chapters.

5

Monolithic Gas Sensor Systems

This chapter includes two different sensor system architectures for monolithic gas sensing systems. Section 5.1 describes a mixed-signal architecture. This is an improved version of the first analog implementation [81,91], which was used to develop a first sensor array (see Sect. 6.1). Based on the experience with these analog devices, a complete sensor system with advanced control, readout and interface circuit was devised. This system includes the circular microhotplate that has been described and characterized in Sect. 4.1. Additionally to the fabrication process, a prototype packaging concept was developed that will be presented in Sect. 5.1.6. A microhotplate with a Pt-temperature sensor requires a different system architecture as will be described in Sect. 5.2. A fully differential analog architecture will be presented, which enables operating temperatures up to 500 °C.

5.1 Single-Ended Mixed-Signal Architecture

5.1.1 System Description

Figure 5.1 shows a block diagram of the single-ended mixed-signal architecture [130–132]. The microhotplate and its components, which were discussed in Sect. 4.1.1, are represented as a single block. The circuitry includes three major components: (1) the metal-oxide resistance readout (Sect. 5.1.3), (2) the microhotplate temperature control loop including a digital proportional-integral-derivative (PID) controller (Sect. 5.1.4), (3) the temperature readout of the bulk chip carrying the electronics (discussed in Sect. 5.1.2), and (4) a standard I^2C serial interface (Sect. 5.1.5). All measured data are read out via the digital interface, which also allows for setting the membrane temperature and the controller parameters.

This microsystem also comprises three 10-bit successive approximation analog-to-digital converters (ADCs) that are used for reading out the microhotplate temperature, the bulk chip temperature, and the sensor resistance, three programmable offset

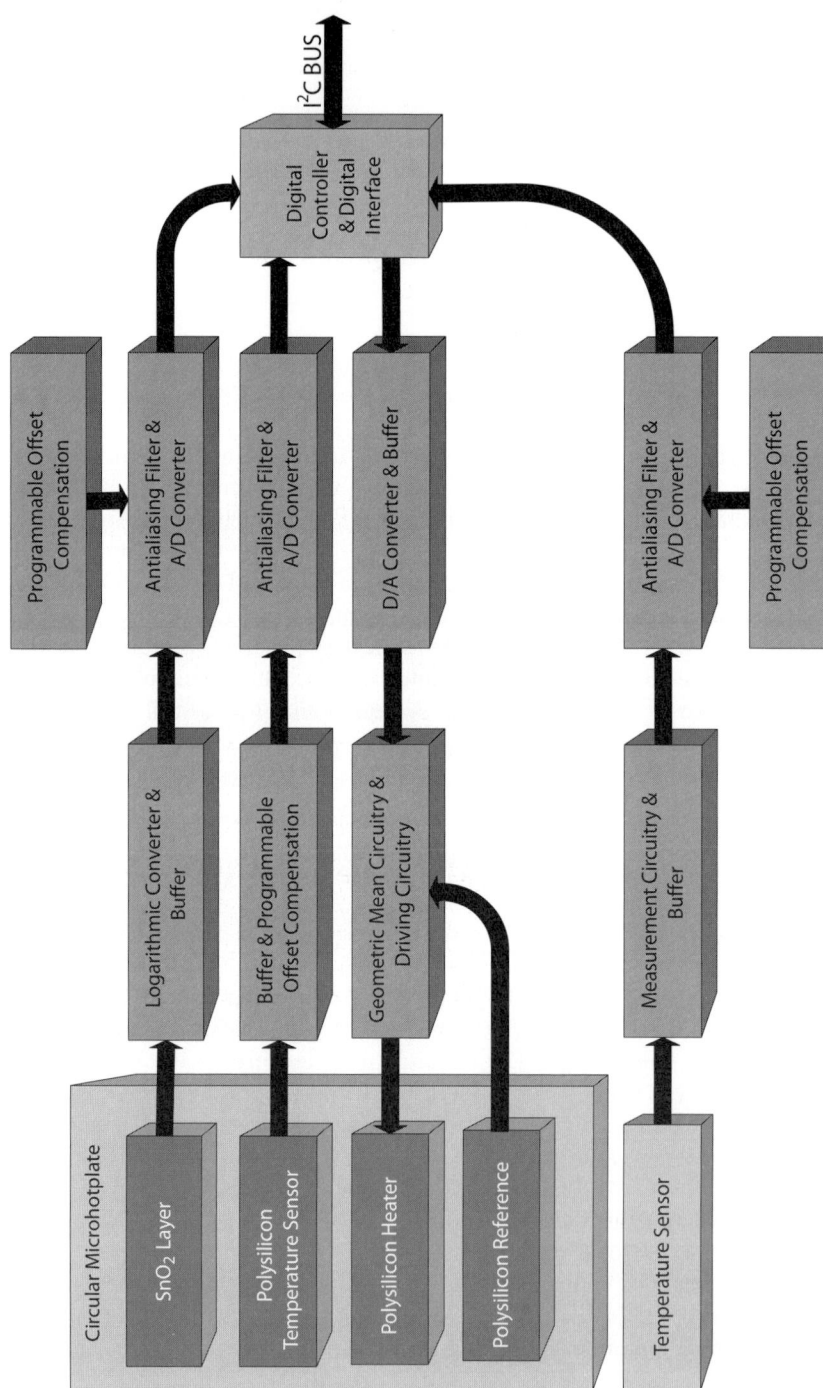

Fig. 5.1. Block diagram of the components of the single-ended mixed-signal architecture

compensators, a flash digital-to-analog converter (DAC), and a bias generator for the reference voltages of the ADCs and the DAC.

The fabrication of the sensor system was described in Sect. 4.1.2, since this microsystem also features a circular microhotplate. A micrograph of the complete microsystem (die size 6.8 × 4.7 mm^2) is shown in Fig. 5.2. The microhotplate is located in the upper section of the chip. The analog circuitry and the A/D and D/A converters are clearly separated and shielded from the digital circuitry. The bulk-chip temperature sensor is located close to the analog circuitry in the center of the chip. The distance between microhotplate and circuitry is comparatively large owing to packaging requirements, as will be explained in Sect. 5.1.6.

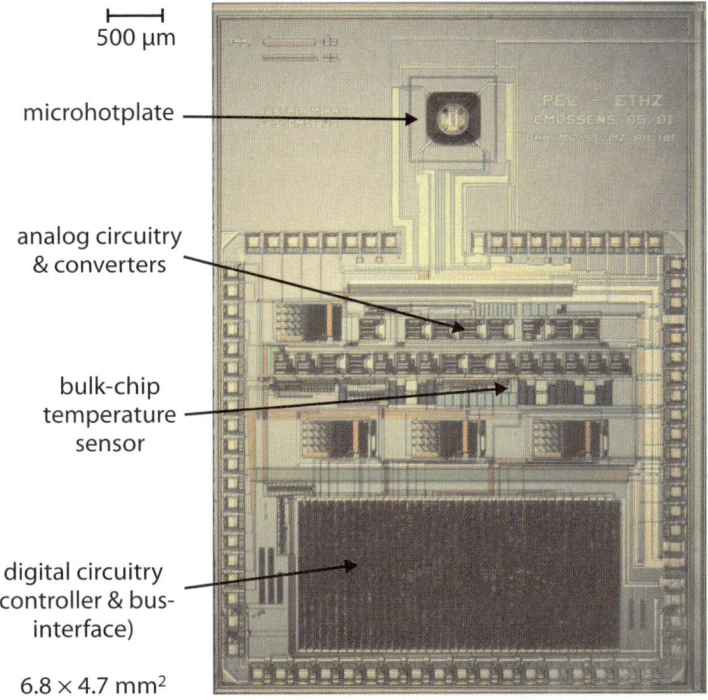

Fig. 5.2. Micrograph of the single-ended mixed-signal microsystem chip

5.1.2 Bulk Chip Temperature Sensor

The schematic of the temperature sensor on the bulk chip is shown in Fig. 5.3. The bulk chip temperature is measured via the voltage difference between a pair of diode-connected pnp-transistors (parasitic transistors as available in the CMOS process, collectors tied to substrate) working at different current densities. Transistor Q_1 is biased with a current of 40 μA, and transistor Q_2 is biased with a current of 10 μA.

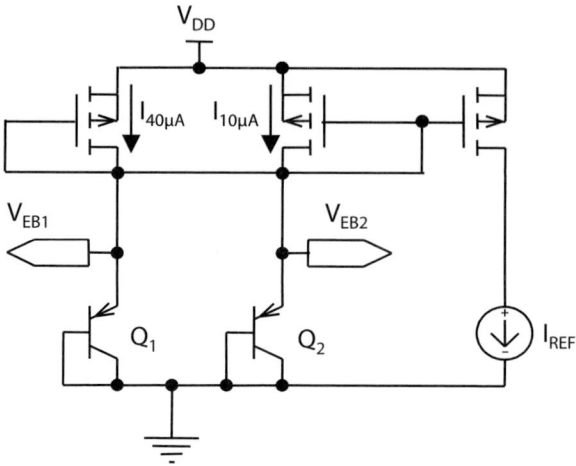

Fig. 5.3. Schematic of the bulk-chip temperature sensor

The voltage difference between the emitter-base voltages (ΔV_{EB}) is proportional to the absolute temperature (T), and is given by Eq. (5.1):

$$\Delta V_{EB} = V_{EB1} - V_{EB2} = \frac{k \cdot T}{q} \cdot \ln\left(\frac{I_{40\,\mu A}}{I_{10\,\mu A}}\right) \tag{5.1}$$

where k is the Boltzmann constant, and q is the electron charge.

The performance of the temperature sensor on the bulk chip is shown in Fig. 5.4. The ambient temperature was swept from -40 to $120\,°C$ in steps of $5\,°C$, and the temperature controllers were switched off. A two-point calibration at $0\,°C$ and $80\,°C$ was performed. The measured sensitivity is about $128\,\mu V/°C$, and the accuracy is better than $1.5\,°C$.

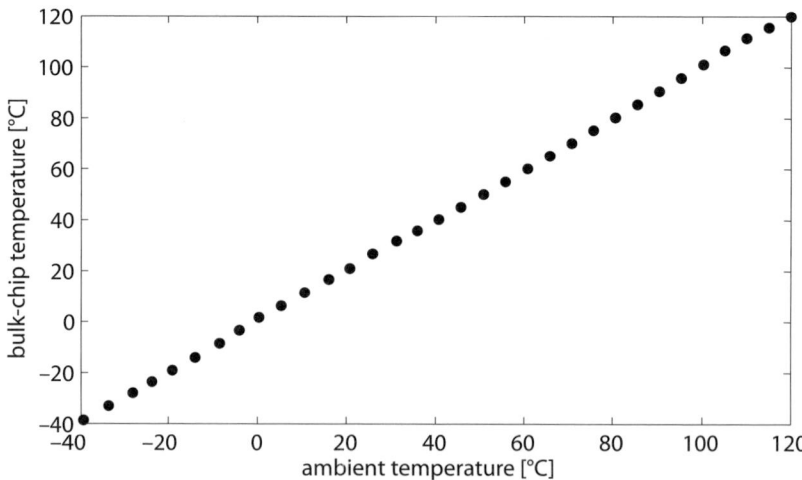

Fig. 5.4. Output of the bulk-chip temperature sensor as a function of the ambient temperature

5.1.3 Logarithmic Converter

The resistance of the tin-oxide layer is read out by a logarithmic converter as a consequence of the large resistance changes between $1\,\text{k}\Omega$ and $10\,\text{M}\Omega$ that are often observed with metal oxides. A logarithmic converter was chosen, since the relation between the metal-oxide resistance and the analyte concentration is nonlinear and can be approximated by a power law (Eq. 2.2). Logarithmic conversion therefore leads to a first-order signal "linearization". The logarithmic converter shown in Fig. 5.5 was designed to cope with this large resistance change.

The logarithmic converter is implemented with a voltage-to-current converter (OPAM, M_1, and R_S) and a pair of diode-connected, vertical pnp transistors (Q_1 and Q_2). The relation between the differential output voltage (ΔV_{EB}) and the sample resistance (R_S) is given by the equation:

$$\Delta V_{EB} = V_{EB1} - V_{EB2} = \frac{k \cdot T_C}{q} \cdot \ln\left(\frac{I_R}{I_{REF}}\right) = -\frac{k \cdot T_C}{q} \cdot \ln\left(R_S \cdot \frac{I_{REF}}{V_{CM}}\right) \qquad (5.2)$$

where k is the Boltzmann constant, q is the electron charge, and T_C is the absolute temperature of the bulk chip.

In the case that $V_{EB} = V_S$, the sensor signal readout of the logarithmic converter, Eq. (5.2) can be simplified to:

$$V_S = -\frac{kT_C}{q} \cdot (\ln(R_S) + c_0) \qquad (5.3)$$

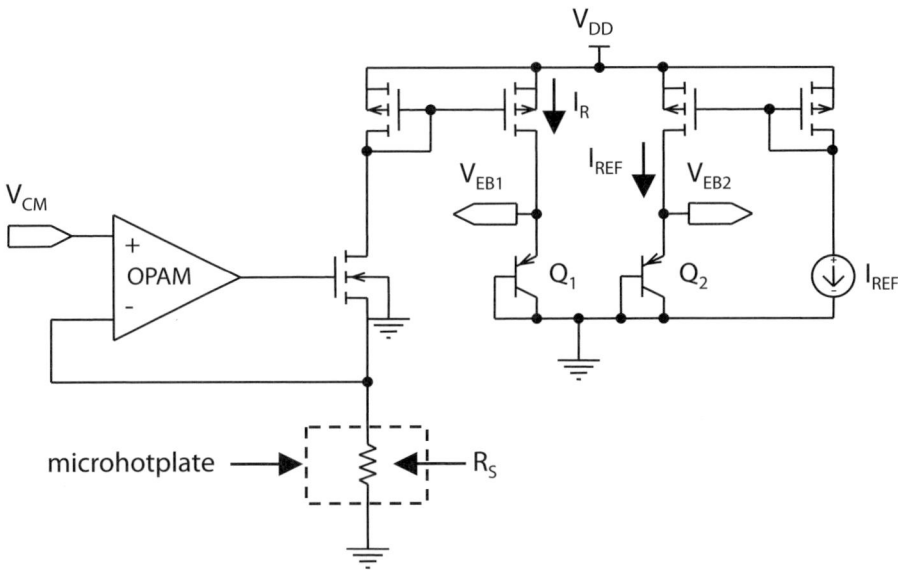

Fig. 5.5. Schematic of the logarithmic converter

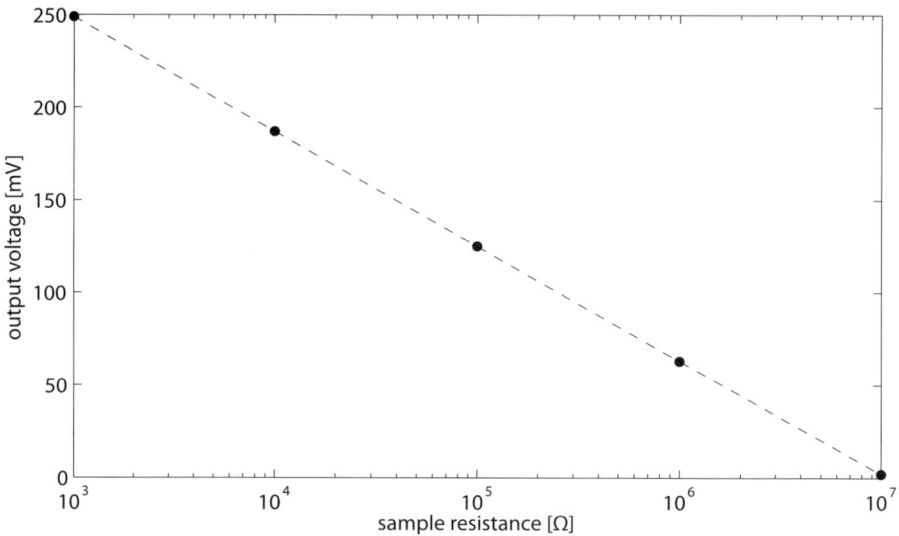

Fig. 5.6. Output voltage of the logarithmic converter as a function of the sample resistance

The negative sign is a consequence of the polarity of the differential output voltage: Lowering the sensor resistance leads to an increased V_S. Equation (5.3) will be used for further discussions of the sensor signal in the section on gas test measurements (Sect. 5.1.9).

The dc level of the logarithmic converter can be changed with the reference current (I_{REF}) or with the common-mode voltage (V_{CM}). The bulk-chip temperature sensor can be used to compensate for the temperature dependence of the logarithmic converter. The performance of the logarithmic converter is shown in Fig. 5.6.

Five calibrated resistors (1% accuracy) with values of $1\,k\Omega$, $10\,k\Omega$, $100\,k\Omega$, $1\,M\Omega$, and $10\,M\Omega$ were connected to the input of the logarithmic converter (i.e., the input, to which the metal-oxide-covered electrodes are connected), and the output voltage (ΔV_{EB}) of the logarithmic converter was measured. The common-mode voltage was $1\,V$, the reference current was $0.1\,\mu A$, and the ambient temperature was kept at $25\,^\circ C$. The offset voltage between the emitter-base voltages was less than $2\,mV$. The resistance values were estimated from the measurements at 2% resolution.

5.1.4 Temperature Control Loop and Geometric Mean Circuitry

The schematic of the temperature control loop and the geometric mean circuitry of the single-ended mixed-signal architecture is shown in Fig. 5.7.

The microhotplate temperature is measured using a polysilicon resistor as temperature sensor (R_T). The resistor is biased with a temperature-independent current source (I_{REF}). The voltage drop across the resistor is low-pass filtered and converted to the digital domain using a 10-bit successive-approximation ADC from the analog library of austria*micro*systems (Unterpremstätten, Austria). The inputs of the digi-

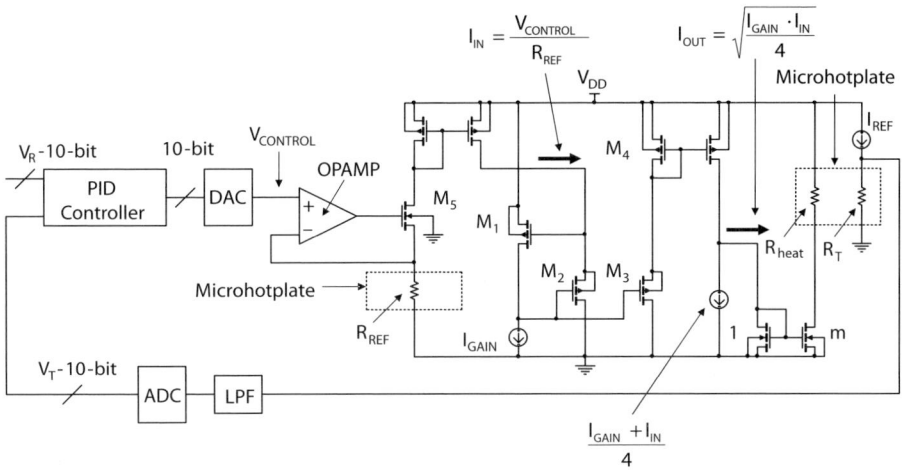

Fig. 5.7. Temperature control loop and geometric mean circuitry of the single-ended mixed-signal architecture

tal PID temperature controller include the feedback signal from the microhotplate temperature (V_T) and the 10-bit reference temperature (V_R). The output of the digital temperature controller is converted to the analog domain using a 10-bit flash DAC ($V_{CONTROL}$). The linearization of the quadratic relationship between $V_{CONTROL}$ and the power dissipated by the polysilicon heater (R_{heat}) facilitates the calculation of the optimal parameters for the digital PID temperature controller. The linearization, which is implemented with a voltage-to-current converter (OPAM, M_5, and R_{REF}) and a geometric mean circuit (transistors $M_1 \ldots M_4$), is described in the following.

The control voltage drives the voltage-to-current converter. The polysilicon reference resistor (R_{REF}) used in the converter is located on the microhotplate in order to be at the same temperature as R_{heat}. This configuration eliminates the temperature dependence of the proportionality coefficient between $V_{CONTROL}$ and the power dissipated by R_{heat} (P_{heat}) as shown in Eq. (5.4):

$$P_{heat} = \left(I_{GAIN} \cdot \frac{m^2}{4} \cdot \frac{R_{heat}}{R_{REF}} \right) \cdot V_{CONTROL} \qquad (5.4)$$

where I_{GAIN} is the bias current of the geometric mean circuit, and m is the transistor ratio of the output current source.

The output current of the voltage-to-current converter is the input current of the geometric mean circuit (I_{IN}). The geometric mean circuit is implemented using a PMOS translinear loop (transistors $M_1 \ldots M_4$) working in the saturation region [137]. The output current of the geometric mean circuit is amplified by a factor m using a current source and is then applied to R_{heat}. The geometric mean circuit can also be used to adjust the open-loop gain of the system by changing the value of the gain current (I_{GAIN}).

The digital PID controller is implemented as a recursive filter with a biquadratic transfer function. Controller wordlength, set-point input, controller output, and controller parameters are 10-bit fixed-point values in the range $[0,1)$. The controller parameters also have an additional sign-bit.

The controller output or actuation signal is clamped in the range $[0,1)$ to avoid overflow, which could render the closed-loop system unstable. The clamping or saturation of the controller output can result in an unacceptable overshooting (known as integral windup) of the controlled value (i.e., the microhotplate temperature) when using integral action. This problem is solved by a reset of the integrator when the controller output is clamped.

A 16-bit programmable frequency divider, which facilitates the selection and scaling of the controller gains, provides the sampling clock of the digital PID controller. The frequency divider has an input clock of 1 MHz.

The corresponding voltage drop across the temperature sensor, $V_T(T_0)$, at a defined ambient temperature, T_0, is initally assessed in calibration measurements. Since the current through the temperature sensor is constant, the relation between measured voltage and resistance of the temperature sensor, R_T, can be written as:

$$\rho = \frac{V_T - V_T(T_0)}{V_T(T_0)} = \frac{R_T - R_T(T_0)}{R_T(T_0)} \tag{5.5}$$

With ρ, the microhotplate temperature, T_M, can be calculated by:

$$T_M - T_0 = \alpha_0 \rho + \alpha_1 \rho^2 \tag{5.6}$$

where α_0 and α_1 are the temperature coefficients of the temperature sensors. These coefficients have to be determined in a calibration procedure, which has been explained in Sect. 4.1.4. The current through the power transistor of the heater driving circuitry is adjusted by the digital controller so that V_T corresponds to a preset reference voltage, V_R. Once the calibration has been done, reversing Eq. (5.6) and replacing V_R for V_T in Eq. (5.5) yields the value of V_R to produce a desired membrane temperature, T_M.

5.1.5 I²C Serial Interface, Instruction Set, and Register Bank

The external communication is handled by an on-chip inter-integrated circuit (I²C) serial interface (Philips, Eindhoven, The Netherlands), which decreases the necessary number of input/output pins for transferring the data from the microsystem. The I²C serial interface is also used for programming the digital controller parameters and some command registers (e.g., programmable sampling rate, programmable analog switches for offset compensation and calibration, etc.).

The I²C header includes the seven-bit address (a_{6-0}) of the chip. Four of these bits a_5–a_2, can be hard-wired during bonding. Internal pull-downs set the default values for these bits to 0.

I²C header

a_6	a_5	a_4	a_3	a_2	a_1	a_0	0

The first bit of the address is designed to be 0 so as to guarantee that I^{2}C arbitration resolves in favor of input messages to the microsystem with respect to output messages. The I^{2}C interface accepts incoming messages of two formats: a simple two-byte command message (A)

Format A

I^{2}C header	Opcode

or a four-byte register-programming message (B), with the 16-bit register value split in its most significant byte (MSB) and least significant byte (LSB)

Format B

I^{2}C header	Opcode	MSB$_{15-8}$	LSB$_{7-0}$

The operation codes or the instruction set are defined as follows (Tab. 5.1):

Table 5.1. Operation codes of the single-ended mixed-signal architecture

Opcode	Binary Format	Description	Format
PWR UP	00000001	Enter full-power mode	A
PWR DN	00000010	Enter low-power mode	A
READ	00000011	Read ADC values	A
STREAM	00000101	Stream (read many) ADC values	A
GET n	$0001n_{3-0}$	Get the value of register n	A
SET n x	$0010n_{3-0}x_{15-0}$	Set the value of register n to x	B

In this microsystem architecture, the I^{2}C recipient is hard-coded to the value 01011000. The I^{2}C sender identifies both, the message format and the address of the chip.

I^{2}C sender

C/$\overline{\text{D}}$	a$_6$	a$_5$	a$_4$	a$_3$	a$_2$	a$_1$	a$_0$

The C/$\overline{\text{D}}$ bit is 1 for a message of format C. The format C is used, when the chip receives a GET command. Then, the chip prompts the I^{2}C interface to write a four-byte message onto the bus, containing the most and least significant bytes of the required register in the MSB and LSB fields, respectively.

Format C

I^{2}C recipient	I^{2}C sender	MSB	LSB

The format D is issued in response to the STREAM and READ opcodes, containing the ten-bit values M, C, S, for microhotplate temperature, chip temperature and sample resistance

Format D

I²C recipient	I²C sender	M_{9-2}	C_{9-2}	S_{9-2}	Mixed bits

The two least significant bits of each value are combined into the *Mixed bits* field.

Mixed bits

0	0	M_1	M_0	C_1	C_0	S_1	S_0

The PWR DN command activates the power-down mode. In this mode, all the analog blocks are switched-off reducing the power consumption from 20 mA (all blocks of the chip are active, microhotplate at 300 °C) to 2 mA (only digital blocks are active).

The register bank of the single-ended mixed-signal architecture is shown in Tab. 5.2.

Table 5.2. Register bank of the single-ended mixed-signal architecture

Word	Bits	Description
0	15–0	*multi-purpose bits:* analog switches for offset compensation, calibration, etc.
1	15–0	delay between streamed samples (in clock cycles) of the I²C interface
2	15–0	sampling period (in clock cycles) of the digital controller
3	9–0	reference or set-point temperature of the digital controller
4–9	10–0	PID controller coefficients: (a_0, a_1, a_2, c, b_1, b_2)

All digital blocks were designed using VHSIC (very high speed integrated circuit) hardware description language (VHDL). The VHDL code was reused for the design of the digital blocks in the microsystems described in Sect. 6.2.

5.1.6 Sensor Packaging

Reliable packaging of chemical microsensors is a challenging task owing to the different requirements for the transducer part (freely accessible with medium or sample contact) and associated electronics (completely protected and shielded), so that only a few prototype packages have been presented so far [23, 140].

Our intention was to find a robust, but simple packaging solution, which relies on known materials for microelectronic packaging. Figure 5.8 shows the packaging concept. A commercially available standard, gold-coated TO-8 socket with 16 pins has been chosen. The chips are affixed with the die-attach EPOTEK H72, which is electrically non-conductive and withstands short-time high-temperature exposure up to 300 °C. The die-attach is only applied underneath the circuitry part, so that the microhotplate cavity is not completely sealed. Pressure, which might build up during heating, can thus be released through the small slit between chip and package. The chip is wire-bonded with a wedge-wedge bonder using Al-wires. To protect the bond

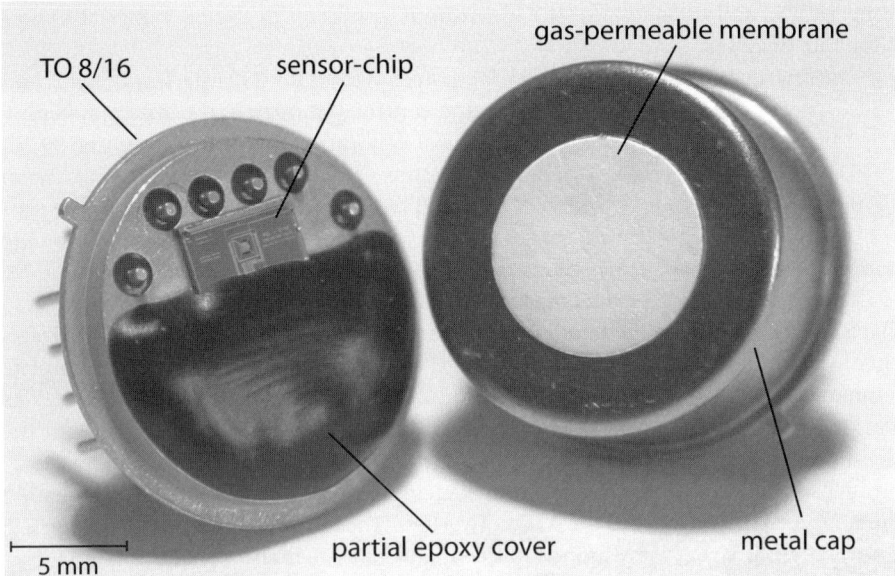

Fig. 5.8. Micrograph of the packaged chip. The *left-hand side* shows the chip attached to a TO8/16 header with a partial epoxy cover. On the *right-hand side*, the metal cap with the gas-permeable membrane is shown

wires and the circuitry, the chip is partially covered with a glob-top epoxy (Loctite HYSOL FP4460). The epoxy is applied with a dispenser and cured for one hour at 125 °C followed by two hours at 160 °C. It is chemically inert and does not affect the sensitive layer during curing. Partial covering of the chips with epoxy allows for a good protection of the electronics, while still enabling free access of the analyte to the sensitive layer and microhotplate. A metal cap with a gas-permeable membrane is mounted on top of the TO-socket (Fig. 5.8) to protect the chip from dust and mechanical impact. The cap can also host metal sieves and filter elements, the latter of which are an important feature in tuning the selectivity of the sensors [5].

5.1.7 Electrial Characterization

A microcontroller board for sensor read-out was developed in collaboration with AppliedSensor (Reutlingen, Germany). This board provides the power supply and the necessary reference voltages and currents to the sensor chip. The sensor chip is mounted on a socket and connected to the microcontroller. The microcontroller provides a set of commands for programming and reading the on-chip registers, controller parameters and sensor values. The microcontroller also manages the data transfer between the I^2C interface of the microchip and a serial RS-232 interface for external read-out via a PC. The data stream is continuously read out in the measurement mode. Each data set contains the temperature of the microhotplate, the chip temperature, and the signal of the logarithmic converter as digital values at a sampling rate of 22 Hz.

The microcontroller board, which is connected to an RS-232 interface of a PC, and a 5.5 V power supply, is all that is needed to control and operate the sensor system.

A two-point bulk-chip temperature sensor calibration was done at 0 °C and 80 °C, by placing the packaged sensor chip in a thermoregulated airstream (ThermoStream TP 04310A, Temptronic, USA) with a Pt-100 temperature sensor affixed to the sensor package. The temperature sensor output voltage is a linear function of the chip temperature within the temperature range of interest (see Eq. (5.1)).

The tracking-mode performance of the digital temperature controller is shown in Fig. 5.9. The microhotplate temperature is plotted versus the digital code of the reference voltage. The 10-bit input range from 0 to 1023 corresponds to a microhotplate temperature range between 170 °C and 310 °C, so that one digit represents 0.1 °C. The tracking error due to noise is less than ± 2 °C. The curve is almost linear with the slight nonlinearity reflecting the second-order coefficient of the polysilicon temperature sensor (see Eq. (5.6)). The controllable temperature range is variable by adjusting the slope, i.e., the step size of the A/D converter of the temperature function in Fig. 5.9. The temperature that corresponds to the digital value 0, is also programmable.

A device with on-chip controller offers several advantages in comparison to conventional microhotplate-based sensors. For uncontrolled microhotplates, the resulting membrane temperature at constant heating power changes due to fluctuations of the gas flow rate, the gas temperature, the ambient temperature, and heat losses as a result of chemical reactions. These temperature variations inevitably produce changes in the sensor signal, since the metal-oxide resistance significantly changes with temperature. In the integrated microsystem, however, temperature fluctuations

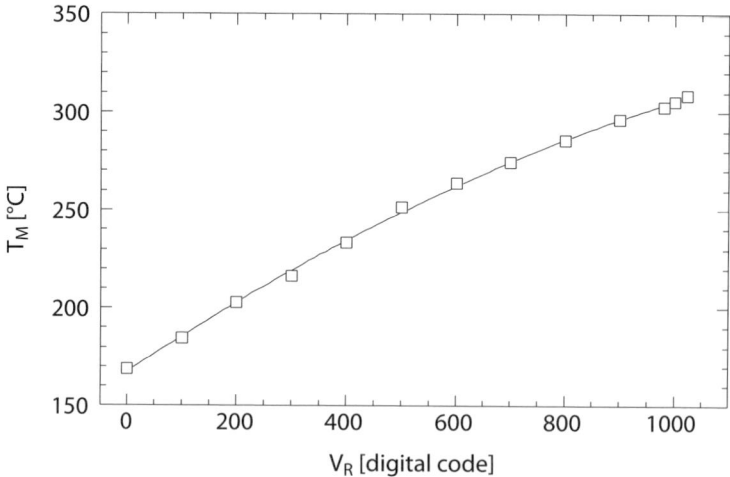

Fig. 5.9. Membrane temperature, T_M, vs. reference voltage, V_R. The reference voltage is given in digital code. The full range of digital values from 1 to 1023 corresponds to a temperature range of 170 °C to 310 °C

are immediately counteracted by the on-chip control circuitry, which keeps a preset temperature constant regardless of the fluctuations in the power consumption. The approximately linear relationship between digital input code and microhotplate temperature directly translates a sinusoidal input signal into a sinusoidal temperature waveform for dynamic measurements. This does not hold for a conventional microhotplate with an uncontrolled resistive heating element. A sinusoidal heating voltage or current waveform does not necessarily produce a sinusoidal temperature modulation as a consequence of the square dependence of the heating power.

The performance of the digital temperature controller in the stabilization mode is shown in Fig. 5.10. The ambient temperature was ramped from −10 °C to 90 °C in steps of 10 °C. A reference-temperature digital code of 500 was programmed, which produced a microhotplate temperature of 250 °C. The microhotplate temperature fluctuations owing to ambient temperature variations are less than ±2 °C.

An important issue of monolithic sensor systems with integrated microhotplates is the overall chip heating. In order to produce a hotplate temperature of 400 °C, the microhotplate and the circuitry typically consume 100 mW, which might cause considerable chip heating. The corresponding chip temperature increase could offset the performance of the on-chip circuitry. The discrepancy between ambient temperature and bulk-silicon chip temperature as a function of the microhotplate temperature in a TO-8 metallic package is shown in Fig. 5.11. The measurement was done at room temperature, and the digital code of the reference temperature was increased in steps of 100 digits. The maximum temperature discrepancy was less than 2 °C. This measurement also shows that the metallic package constitutes a better heat sink than the ceramic package.

Such slight chip temperature increases do definitely not affect the on-chip circuitry, which is qualified for temperatures between −40 °C and 80 °C. Chip heating

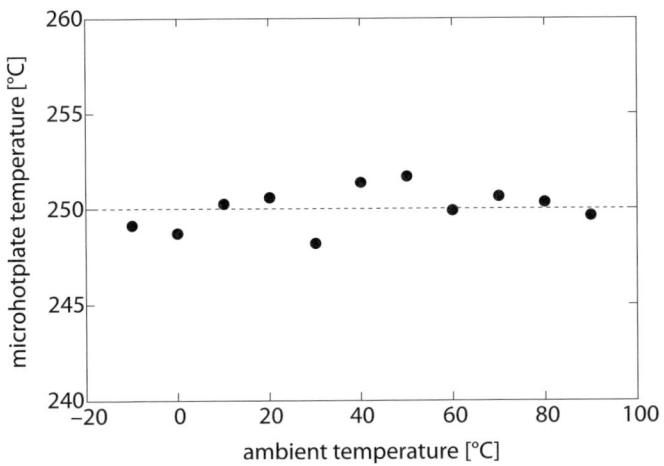

Fig. 5.10. Performance of the digital PID temperature controller of the single-ended mixed-signal architecture in the stabilization mode

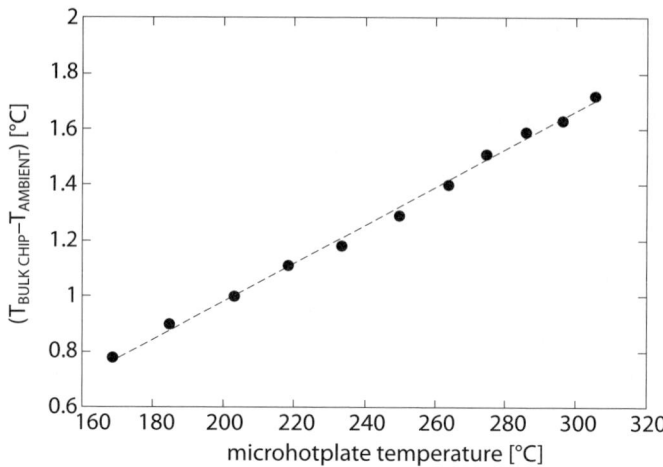

Fig. 5.11. Discrepancy between ambient temperature and bulk-chip temperature as a function of the microhotplate temperature for the single-ended mixed-signal architecture (TO-8 package)

can be further minimized by specific packaging methods, such as the use of die-attach materials with high thermal conductivity.

5.1.8 Gas Test Measurement Setup

Gas test measurements were performed in a gas manifold. Vapors were generated from gas bottles with calibrated target gas concentrations (e.g. CO) and, then, diluted as desired using computer-driven mass-flow controllers and synthetic air (oxygen/nitrogen mixture without humidity) as carrier gas. The different humidity contents were generated by passing a fraction of the overall gas stream through water bubblers, which were kept at constant temperature. All vapors were mixed and temperature-stabilized before entering the thermoregulated test chamber. The gas tubing in the manifold was made of stainless steel. The sensors were mounted inside a flow-through stainless steel cell, and the measurements were performed at 30 °C chamber or chip temperature. The thermostat used for the measuring chamber was a microprocessor-controlled Julabo FP 30 MH (Julabo, Seelbach, Germany). The gas flow-rate to the sensors was 200 sccm at a total pressure of 10^5 Pa. The response time of the sensors is on the order of seconds (< 10 s). It takes, however, approximately 1 minute to reach an equilibrium state in the set-up. This time span is needed to achieve a constant gas concentration (steady state) in the chamber (volume 10 ml) at the chosen flow rate. Typical experiments consisted of alternating exposures to pure synthetic air and analyte-loaded air. The analyte gas concentrations in the measurement chamber were routinely checked using a photoacoustic instrument (Innova Airtech, Ballerup, Denmark).

5.1.9 Gas Test Measurements

A series of chemical measurements was carried out in order to test the sensor performance. The microhotplate was heated to a defined temperature, the sensor was exposed to the analyte gas at a certain concentration, and the resulting sensor signal was recorded. Carbon monoxide was used as a test analyte, and nanocrystalline SnO_2 with 0.2 wt% Pd doping was used as sensitive layer, which is optimized for CO detection. The sensor signal, S, is expressed in digital units after the conversion of V_S (see Eq. (5.3)) by the on-chip A/D converter. The raw data (digital numbers in the 10-bit output range) were processed with a moving-average filter averaging over 100 data points. Figure 5.12 shows a typical sensor response. The sensor signal output range from 200 to 280 digital units corresponds to a resistance range of 250 to 50 kΩ with higher digital values representing lower resistance. Consequently, the resistance drop upon the presence of CO, as a reducing gas leads to higher sensor readings. Each analyte exposure step is 15 min at a constant relative humidity of 40% (23 °C humidifier temperature) and 30 °C chip or ambient temperature. The microhotplate operating temperature was 275 °C. Low CO-concentrations of 1, 3 and 5 ppm (partial pressure 0.1–0.5 Pa CO) were dosed to the sensor. As can be seen in Fig. 5.12, a concentration of 1 ppm CO is clearly detectable.

The noise in the sensor signal is ±0.5 digits. This corresponds to a 1-digit quantization noise of the A/D converter. At low concentration levels, the noise is equivalent to a CO-concentration variation of ±0.03 ppm. Multiplying this value by three, the limit of detection was assessed to be 0.1 ppm. The gas concentration resolution amounts

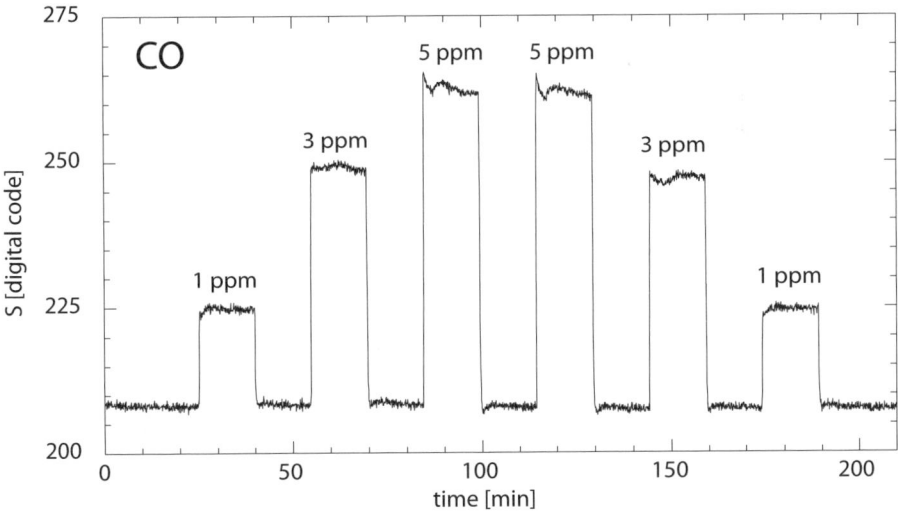

Fig. 5.12. Sensor responses upon exposure to different concentrations of CO at a hotplate temperature of 275 °C and 40% relative humidity. A higher sensor signal corresponds to a lower resistance value of the sensitive layer. The digital sensor output from 200 to 280 covers a resistance range from 250 to 50 kΩ

to ±0.2 ppm. The sensor baseline, S_{air}, represents the sensor reading in humidified synthetic air without any analyte present. The difference between this baseline value before the analyte dosing onset and the sensor signal, S, before the analyte dosing end, was used to determine the sensor response at a given concentration. The full digital output range of S from 0 to 1023 covers sensor resistances from 10 MΩ to 1 kΩ. Taking into account the A/D-conversion of the sensor readout value, S, and applying Eq. (5.3), ΔS can be expressed as:

$$\Delta S = S - S_{\text{air}} = -\left(c_S \cdot \ln\left(R_S\right) - c_S \cdot \ln\left(R_S^{\text{air}}\right)\right) = c_S \cdot \ln\left(\frac{R_S^{\text{air}}}{R_S}\right) \qquad (5.7)$$

The proportional constant, c_S, includes the conversion relation of the A/D converter and the factor kT_C/q of Eq. (5.3). Since the measurement chamber is temperature-stabilized, the chip temperature, T_C, is constant. R_S^{air} is the resistance of the sensitive layer in synthetic air, and R_S is the resistance upon analyte exposure. Equation (5.7) relates ΔS to the ratio R_S^{air}/R_S, which is commonly plotted as sensor signal for reducing gases.

The sensor response at a defined relative humidity of 40% (23 °C humidifier temperature) is displayed in Fig. 5.13 for various microhotplate temperatures. ΔS in-

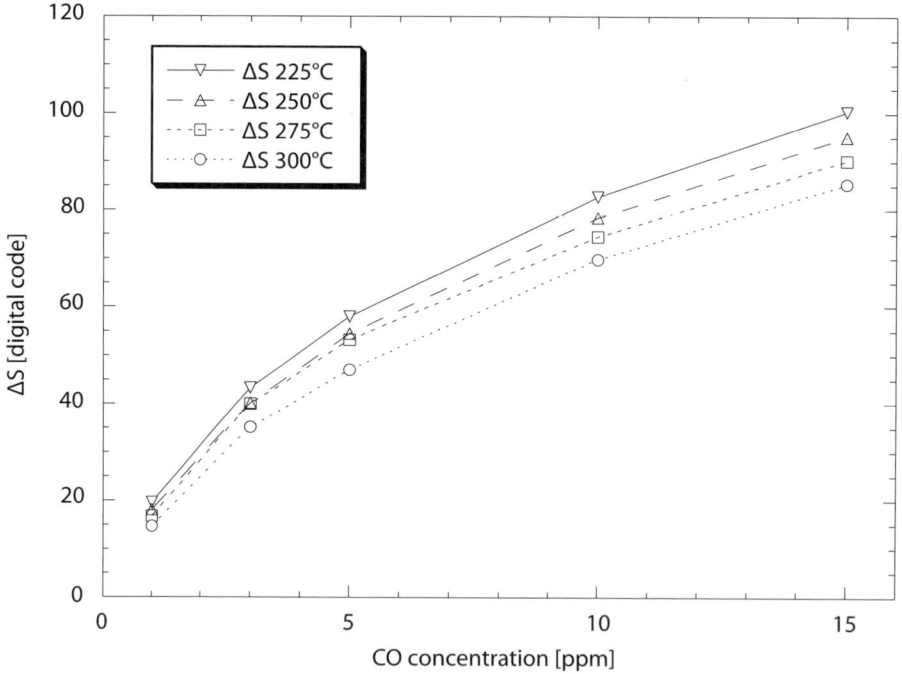

Fig. 5.13. Sensor response, ΔS, upon exposure to different CO concentrations. The humidity was kept constantly at 40% r.h. The graphs have been recorded at different microhotplate temperatures (225–300 °C)

creases with decreasing microhotplate temperature upon dosage of different CO concentrations. The highest sensitivity was achieved for the lowest operation temperature of 225 °C. The sensor signal is hence strongly temperature-dependent with the highest sensitivity being achieved at the lowest operating temperature of 225 °C. The temperature dependence for ΔS in the range between 250 °C and 300 °C is 0.15 digits per °C at a CO concentration of 5 ppm. A temperature change of 1 °C at 5 ppm CO and at a microhotplate temperature of 275 °C thus produces the same signal as a CO concentration change of 0.04 ppm. As mentioned previously, the resolution of the temperature controller was estimated to be ±2 °C, which corresponds to an uncertainty of ±0.1 ppm for the CO measurement. As this is in the same order of magnitude as the resolution in the concentration measurements, it is evident from these considerations, that a precise temperature control is absolutely indispensable to advance into low-concentration and threshold-level measurements.

In order to assess the dependence of the output signal on changes in the humidity content of the sample gas, an additional series of measurements was carried out. The hotplate temperature was set to 275 °C, and CO measurements were recorded at three different humidity levels (10, 20 and 40% r.h.) The humidifier temperature was set to 23 °C, and the chip temperature was 30 °C. As can be seen in Fig. 5.14, the sensor response increases with increasing humidity. The large sensor response difference between 10% and 20% r.h. shows that this effect is more pronounced at low humid-

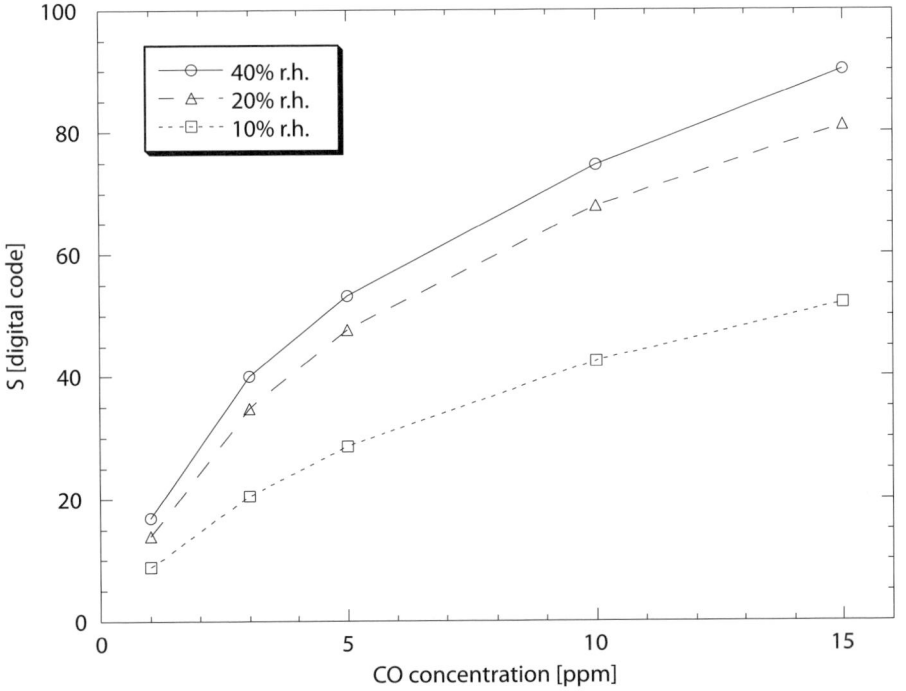

Fig. 5.14. Sensor responses, ΔS, versus CO concentration at different humidity levels (10–40%)

ity levels in comparison to higher humidity levels. These results were expected and reflect the well-established role of water in the CO-sensing mechanism for tin-oxide layers (Sect. 2.3.1).

5.2 Differential Analog Architecture for Operating Temperatures up to 500 °C

5.2.1 System Description

The limit for the operating temperature of CMOS-microhotplates can be extended by using the microhotplate that was presented in Sect. 4.3. We now detail high-temperature microhotplates with Pt-resistors that have been realized as a single-chip device with integrated circuitry. While the aluminum-based devices presented in Sect. 4.1 were limited to 350 °C, these improved microhotplates can be heated to temperatures up to 500 °C. As the typical resistance value of the Pt-resistor is between 50 and 100 Ω, a chip architecture adapted to the low temperature sensor resistance was developed. The system performance was assessed, and chemical measurements have been performed that demonstrate the full functionality of the chip.

The system architecture is shown in the block diagram of the microsystem in Fig. 5.15. The microhotplate hosts the SnO_2 sensing resistor, the Pt temperature sensor resistor and a polysilicon-heating resistor. The layout of these elements and the microhotplate were already described in Sect. 4.3. The SnO_2 sensing resistor is connected to the logarithmic converter, which features a programmable input range, in addition to the functions described in Sect. 5.1. The range is selected by means of switches that determine the input currents to the logarithmic converters, so that it is possible to adapt

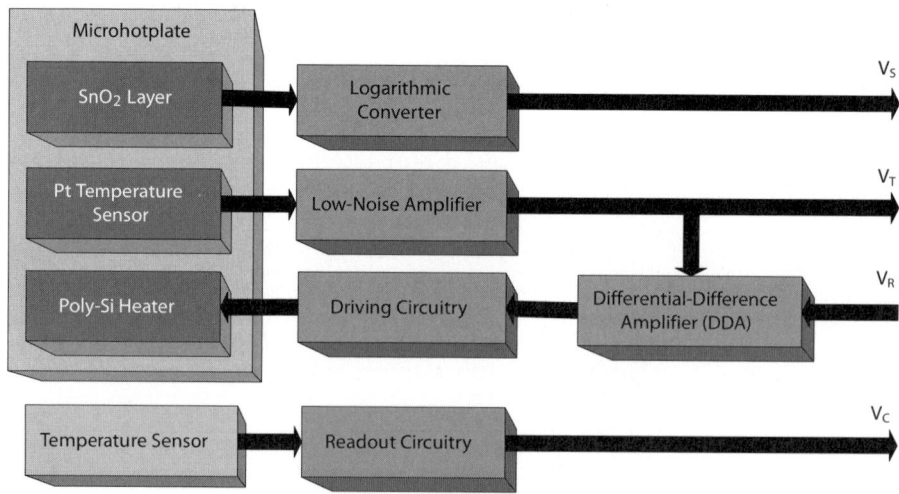

Fig. 5.15. Block diagram of the differential analog system architecture for high operation temperatures

the input range to the expected resistance range of the sensitive layer. Instead of using an on-chip A/D-converter, the output voltage of the converter is directly read out by a multimeter. The temperature sensor is driven by an on-chip constant-current source with a current of 500 μA. As the Pt-resistor used as temperature sensor has a comparatively low resistance (typically 75 Ω), the resulting temperature-dependent voltage drop has to be amplified by a low-noise amplifier. The output voltage of the amplifier is filtered by a low-pass filter and serves as feedback signal for the temperature controller. An analog-proportional temperature controller was implemented using a differential-difference amplifier (DDA). The DDA is connected to a power transistor, which drives the heating resistor with a nominal value of 200 Ω. The current through the transistor is limited by a degeneration resistor of 50 Ω. The current limiter enhances the long-term reliability, because current overshoots are avoided, which may occur, e.g., in case of very rapid changes at the control voltage input. Higher microhotplate temperatures can be achieved by reducing the resistance of the degeneration resistor through connecting additional resistors in parallel or through introducing a shortcut. Finally, the bulk temperature sensor is implemented by a pair of diode-connected PNP-transistors (Sect. 5.1.2).

A micrograph of the sensor chip is shown in Fig. 5.16. The microhotplate is located in the chip center with the circuitry in the lower section of the chip. After the post-CMOS micromachining steps, the chips were affixed to a DIL package and bonded.

microhotplate

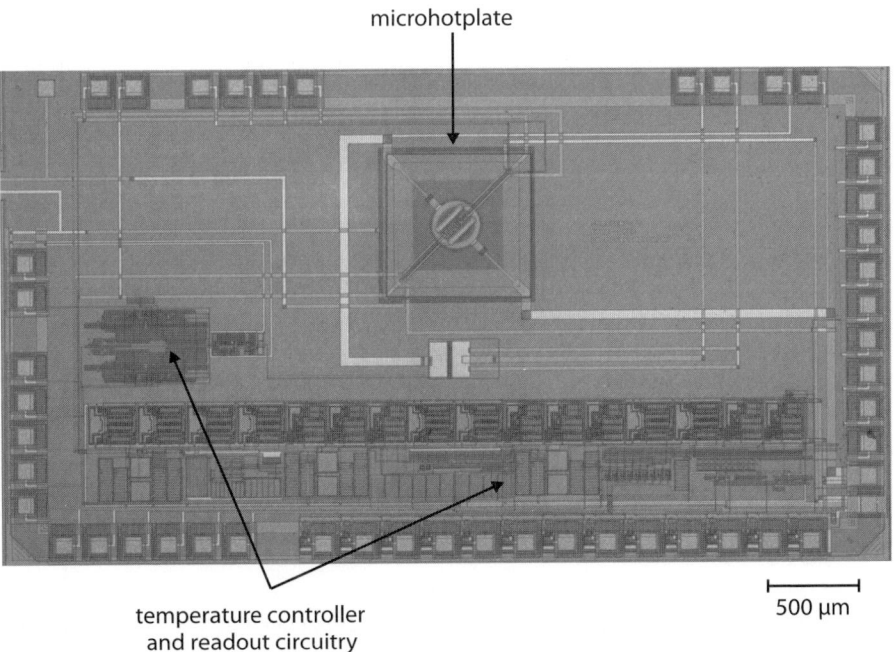

temperature controller
and readout circuitry

500 μm

Fig. 5.16. Micrograph of the sensor microsystem with integrated Pt temperature sensor for operating temperatures up to 500 °C

As the chip layout is not particularly optimized for a TO-package, a DIL package was used. In contrast to the previous sensor fabrication process, the sensitive layer was deposited after packaging, and an on-chip annealing was performed.

5.2.2 Temperature Control Loop

The schematic of the temperature control loop is shown in Fig. 5.17.

The temperature on the microhotplate is controlled between room temperature and $500\,^{\circ}$C, and is measured using a platinum resistor as temperature sensor (R_{Pt_T}) with a nominal resistance of 75 Ω. The resistor is biased from a temperature-independent current source (I_{BIAS}). The voltage drop across the platinum resistor is amplified by a fully-differential low-noise amplifier (LNA) that will be described later in this section. The amplified differential signal is low-pass (LPF) filtered and provides the feedback signal to the differential analog proportional microhotplate temperature controller, which is implemented by a rail-to-rail differential-difference amplifier (DDA). The DDA drives the power transistor (M_1). The power transistor can be switched off for recalibration of the temperature sensor using the power-down transistor (M_2). The power transistor also has a degeneration resistor (R_{DEG}) at the source, which limits the drain current. The differential architecture of this temperature control loop improves the power supply rejection ratio.

The dominant pole of this temperature control system is also determined by the thermal time constant of the microhotplate, which is approximately 20 ms. The open-loop gain of the differential analog architecture (A_{OL_DAA}) is given by Eq. (5.8):

$$A_{OL\_DAA} = A_{OL\_DDA} \cdot \frac{R_{heat}}{R_{DEG}} \tag{5.8}$$

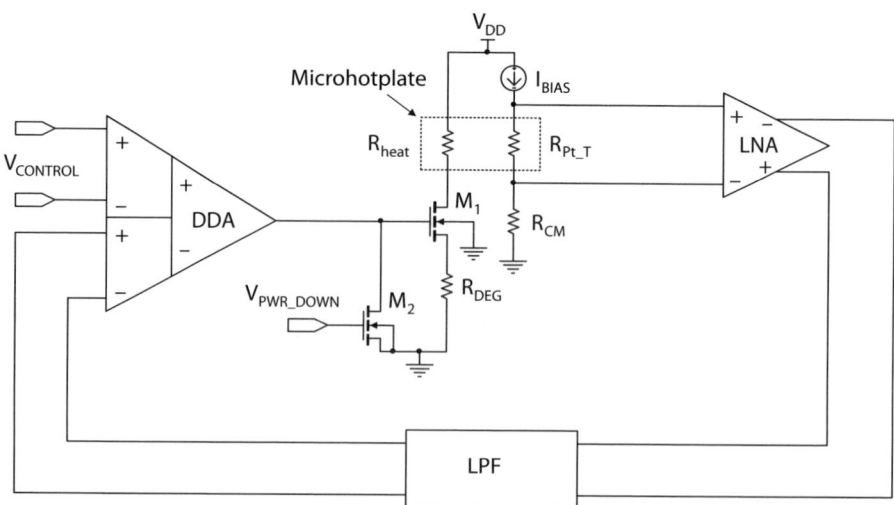

Fig. 5.17. Temperature control loop of the differential analog architecture

Low-noise amplifier

The voltage drop across the platinum temperature sensor is small since the platinum resistor has a nominal resistance of only 75 Ω. The fully-differential LNA amplifies the minute voltage drop in order to provide an useful feedback signal to the differential-analog proportional controller. A simplified schematic of the fully-differential low-noise amplifier is shown in Fig. 5.18.

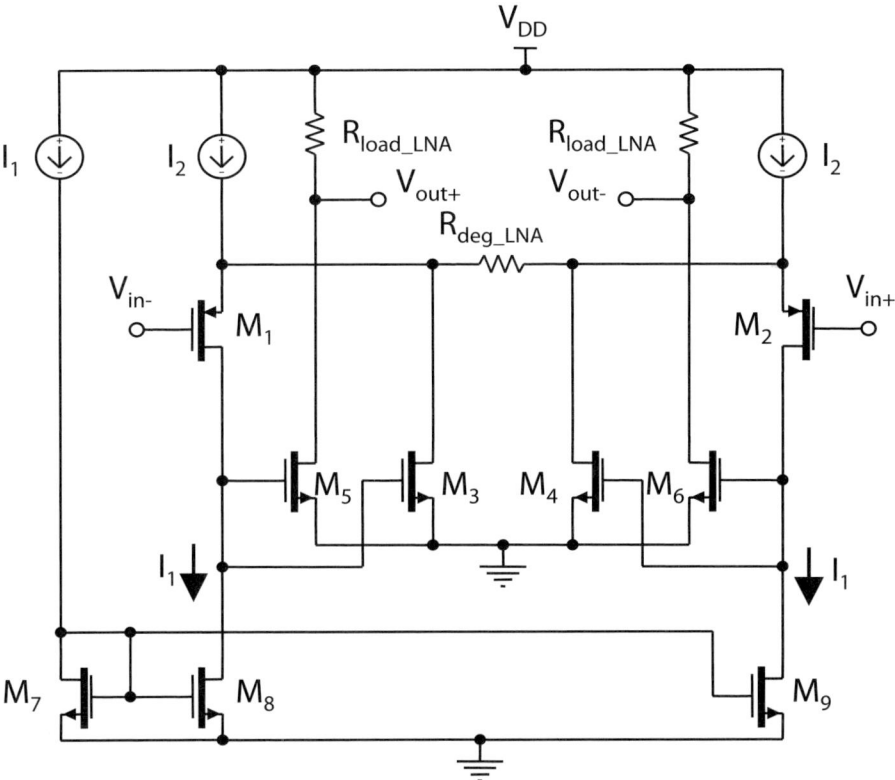

Fig. 5.18. Simplified schematic of the fully differential low-noise amplifier

The differential input voltage ($v_{in} = V_{in}^+ - V_{in}^-$) is copied over the series impedance of the transconductance of the input transistor pair (M_1 and M_2) and the degeneration resistor (R_{deg_LNA}). With the transistors M_3 and M_4 acting as gain stages, the input transistor pair is forced to conduct the static current I_1. These input transistors ideally act as dc level shifters, creating a linear copy of the input signals from the gate to the source. The gain transistors conduct a dc current ($I_2 - I_1$) plus the ac current (i_{ac}), which is determined by the differential voltage input (v_{in}) and the degeneration resistor. The ac current (i_{ac}) is copied by transistors M_5 and M_6, and then drives the load resistors (R_{load_LNA}). The gain of the low-noise amplifier is given by Eq. (5.9):

$$A_{\text{LNA}} \cong \frac{2 \cdot R_{\text{load\_LNA}}}{R_{\text{deg\_LNA}}} \qquad (5.9)$$

Transistors M_8 and M_9 are the main sources of noise of this amplifier. Their area is hence optimized in order to meet the noise specifications. Table 5.3 summarizes the transistor dimensions, bias currents and resistance values.

Table 5.3. Transistor dimensions, bias currents, and resistance values of the fully differential low-noise amplifier

Transistors	Dimensions (W[μm]/L[μm])
M_1, M_2	200/10
M_3, M_4, M_5, M_6	150/20
M_7, M_8, M_9	200/30

Currents	Value [μA]
I_1	40
I_2	80

Resistors	Value [kΩ]
$R_{\text{deg\_LNA}}$	6
$R_{\text{load\_LNA}}$	64

The experimental results of the fully differential low-noise amplifier are summarized in Tab. 5.4.

Table 5.4. Experimental characterization of the fully differential low-noise amplifier

Parameter	Value
Power supply	5 V
Current consumption	320 μA
Gain	26 dB
Input-referred noise (0–30 kHz)	<7 μV
Unity-gain bandwidth ($C_L = 2$pF)	5 MHz
Input-referred offset	<5 mV
Linear input range	$0.5\,V < V_{\text{in}} < 4.5\,V$
Technology	0.8 μm CMOS

5.2.3 Sensor Measurements

For the characterization of the sensor system, the voltage drop across the Pt-resistor and the chip-temperature sensor have been read out with multimeters. The input control voltage is set using a high-precision universal source. A set of measurements was carried out with unheated membranes at ambient temperature to determine the initial voltage drop over the Pt-sensor and the bulk temperature sensor. The temperature for the respective output voltage, V_T, of the temperature sensor of the heated microhotplate can be determined by using the calibration polynomial. The input control voltage, V_R, was increased in steps of 50 mV. The control voltage range from 0.7 to 0.95 V in Fig. 5.19 corresponds to a microhotplate temperature range from ambient temperature to 500 °C, which clearly exceeds the CMOS temperature limit of 350 °C. The controller showed a resolution of ±2 °C.

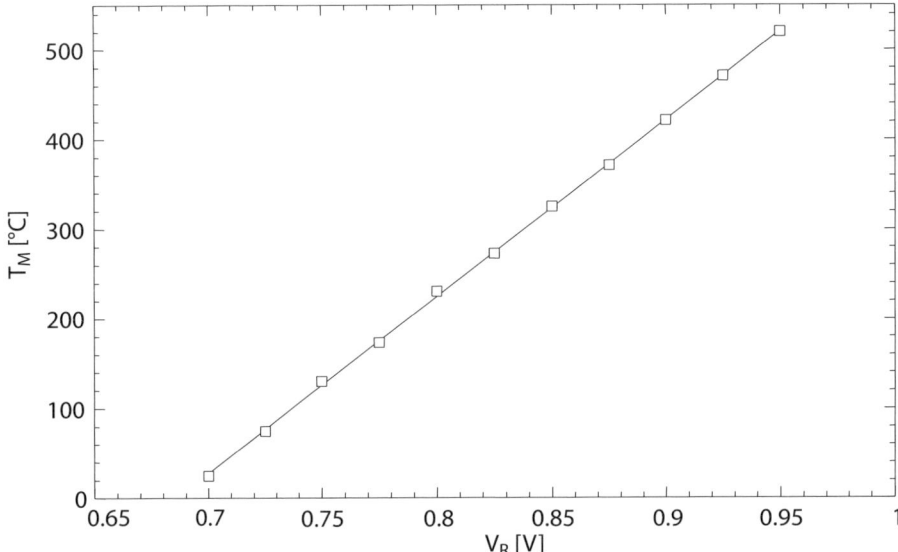

Fig. 5.19. Tracking mode operation performance of the temperature controller

The relation between input voltage and microhotplate temperature is linear, which is a result of the good linearity of the Pt-sensor. Such transfer characteristics are advantageous in case of input modulation and for potential recalibration.

The temperature changes of the bulk chip upon microhotplate heating were assessed. The chip was mounted in a standard ceramic DIL package. The discrepancy between ambient temperature and the bulk-silicon chip temperature was measured as a function of the microhotplate temperature and is shown in Fig. 5.20. The measurement was done at room temperature, and the control voltage was increased in steps of 25 mV thus heating the membrane from room temperature to 500 °C. The maximum discrepancy between bulk chip temperature and ambient temperature was less than 4 °C, which demonstrates the excellent thermal isolation between the microhotplate on the dielectric membrane and the bulk substrate.

Chemical measurements were performed in order to show the full functionality of the chip. Because this chip had been bonded first and then coated, on-chip annealing of the deposited material was successfully performed for one hour at 400 °C. The sensitive layer was heated to 290 °C in the chemical measurement setup and exposed to different concentrations of CO at a relative humidity of 40%. The output voltage of

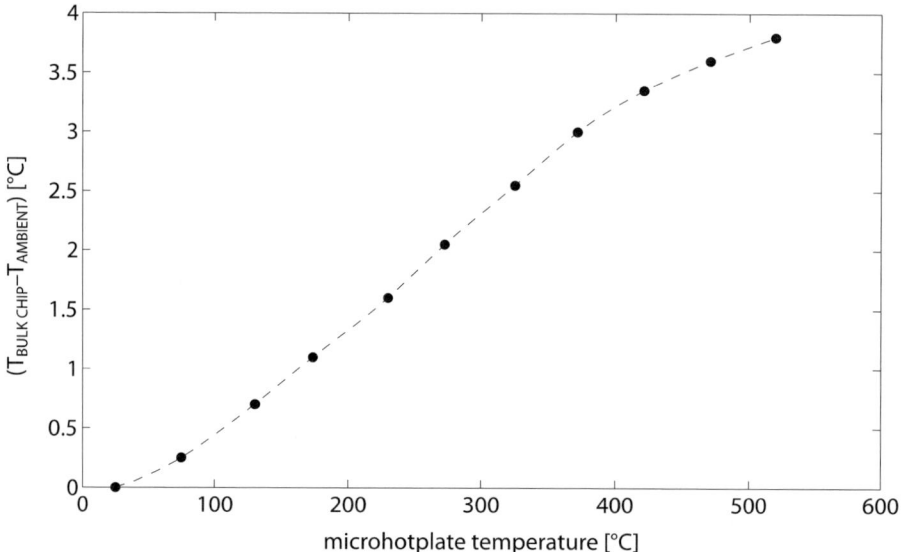

Fig. 5.20. Discrepancy between ambient temperature and bulk-chip temperature as a function of the microhotplate temperature for the differential analog system chip (DIL package)

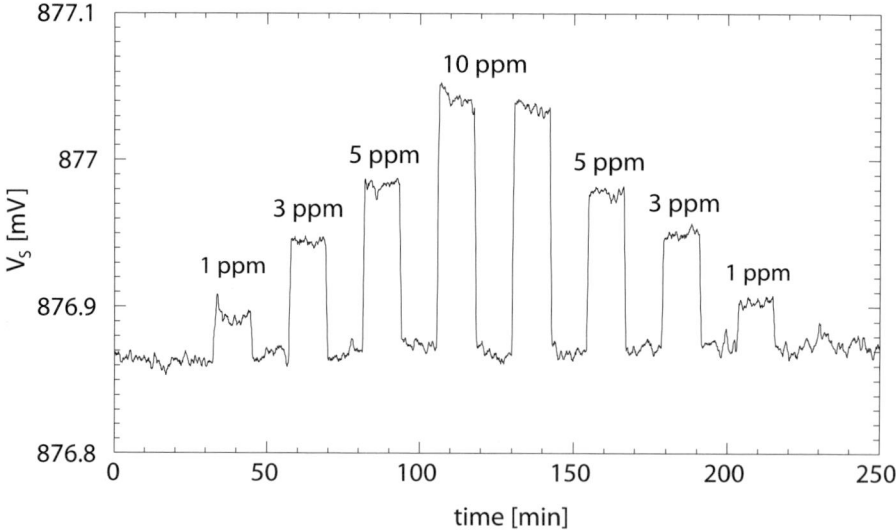

Fig. 5.21. Sensor response of a sensor system featuring a Pt-temperature sensor upon exposure to CO. The microhotplate temperature was 290 °C, and the measurements were conducted at 40% r.h.

the logarithmic converter was directly measured using a multimeter (Fig. 5.21). The data were processed using a moving-average filter including 10 measurement points in order to reduce the noise. A higher signal reading in V_S corresponds, again, to a decrease in the sensor resistance.

A concentration of 1 ppm of CO is clearly detectable. The standard deviation in a signal plateau corresponds to ±0.3 ppm. For a differential sensor signal, a resolution of ±0.5 ppm for a concentration of 5 ppm was measured, which shows that satisfactory metal oxide sensitivity can also be achieved with on-chip annealing.

In conclusion, the full functionality of the integrated sensor chip was demonstrated. The possibility of using higher operating and annealing temperatures renders the system suitable for investigations on a large variety of sensing materials. The system is also suitable for CH_4-sensing with a SnO_2-thick film, where operating temperatures of more than 350 °C are commonly applied.

6

Microsensor Arrays

A well-known problem of tin dioxide is its lack of selectivity (Chap. 2). This situation is usually dealt with by using an array of sensors in combination with multi-component or pattern recognition algorithms such as principal component regression (PCR), multi-way analysis or artificial neural networks (ANN) [142, 143]. Doping of the tin dioxide also changes its selectivity characteristics to different gases [68]. Another parameter that can be varied is the operation temperature. The use of an array of microhotplates with individually controlled temperatures, the hotplates of which are covered with different sensitive materials, increases the overall information that can be extracted from metal-oxide-based gas sensing systems.

While in the previous chapters, only sensor systems featuring a single sensor were discussed, we here present different microsensor systems featuring arrays of three microhotplates. In all these systems each individual microhotplate has its own on-chip temperature controller that provides individual thermoregulation. Thus, the three microhotplates can be coated with three different sensing materials, and each material can be operated at its optimum temperature. Moreover, the application of any kind of temperature program to the hotplates becomes possible. The first system described in Sect. 6.1 features a single-ended analog architecture, which is the simplest implementation of such a sensor array. The second array chip (Sect. 6.2) is a realization of a differential mixed-signal architecture. This system is an extension of the system presented in Sect. 5.1. However, the microhotplate temperature sensor is read out in a differential configuration, and the mean square-circuit in the individual temperature-control loop has been removed. The most advanced system will be presented in Sect. 6.3 and follows a different philosophy. An architecture with MOSFET-heated microhotplates that have been integrated in a fully digital temperature control loop has been realized.

6.1 Single-Ended Analog Architecture

The singled-ended analog architecture comprises a temperature sensor on the bulk chip, three single-ended analog proportional microhotplate temperature controllers (one controller per microhotplate), and an array of three circular microhotplates, each equipped with a polysilicon resistor as temperature sensor, which enables controlling the operation temperature from room temperature up to 350 °C [144, 145]. A simplified block diagram of the singled-ended analog architecture is shown in Fig. 6.1. The resistance of the tin-oxide layer is directly accessible and read out by a multimeter. Since this microsystem includes the circular microhotplate, the fabrication of the sensor array is analogous to the fabrication sequence as described in Sect. 4.1.2.

A micrograph of the single-ended hotplate-based microsystem is shown in Fig. 6.2 and features a die size of 5.0×2.9 mm^2. This system is a minimal implementation of a temperature-controlled microhotplate system. Temperature modulation is facilitated by an direct access to the input voltage: A modulation of the input voltage is translated into a modulation of the microhotplate temperature. Another interesting application of the system includes its use as a microcalorimeter or as a material research platform [145]. The schematic of the temperature-control loop is shown in Fig. 6.3.

The single-ended proportional controller was implemented using an operational amplifier (OPAM) with source degeneration [133–136]. The operational amplifier drives a power transistor (M_1), which provides the current to the polysilicon heater (R_{heat}, 125 Ω nominal). The power transistor has a polysilicon degeneration resistor (R_{DEG}) at the source with a nominal resistance value of 50 Ω. The degeneration resistor limits the drain current of the power transistor, which enhances the long-term reliability, since any potential electromigration at the microhotplate aluminum connections at the applied high temperatures will be reduced.

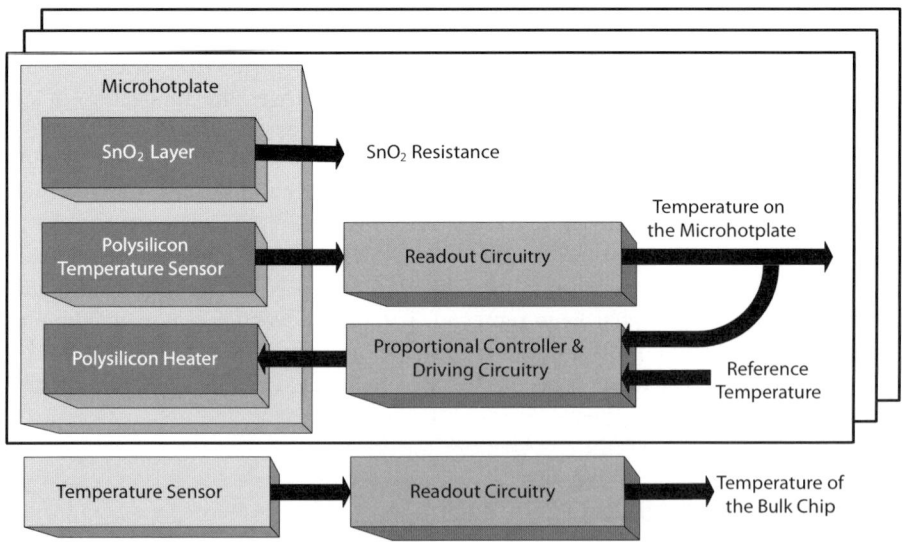

Fig. 6.1. Block diagram of the single-ended analog architecture

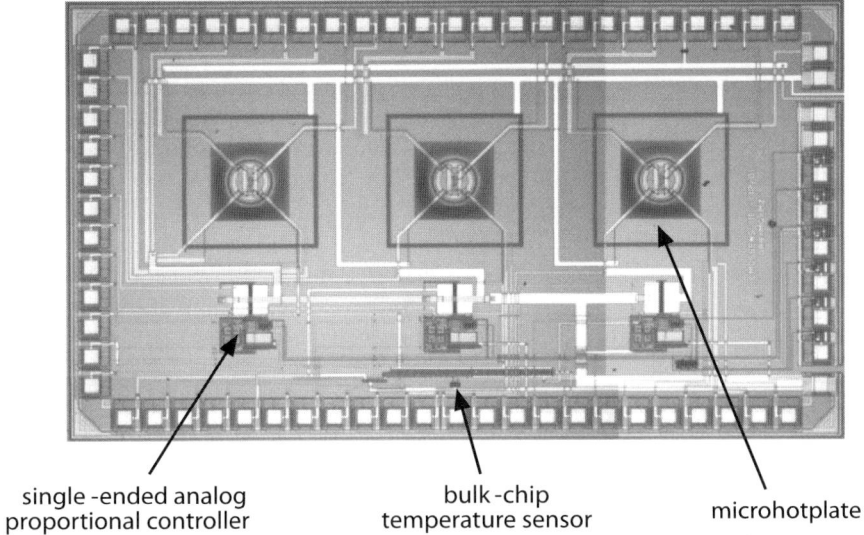

single -ended analog   bulk -chip
proportional controller   temperature sensor   microhotplate

Fig. 6.2. Micrograph of the single-ended analog hotplate-based microsystem

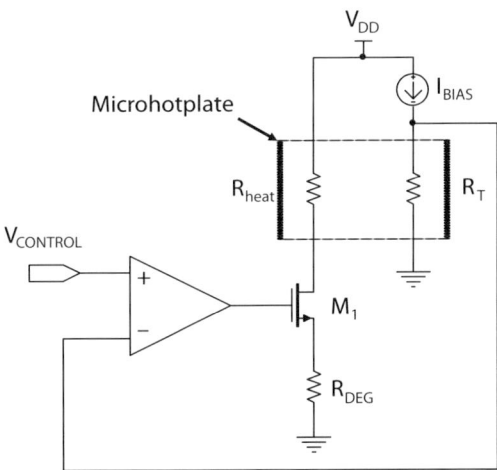

Fig. 6.3. Temperature-control loop of the single-ended analog architecture

The temperature sensor is located in the microhotplate center (R_T, 10 kΩ nominal). This polysilicon resistor is biased with a temperature-independent current source (I_{BIAS}). The voltage-drop across the polysilicon temperature sensor provides the feedback signal for the temperature controller.

The dominant pole of this temperature control system is determined by the thermal time constant of the microhotplate, which is approximately 20 ms. The open-loop gain of the single-ended analog architecture (A_{OL_SEAA}) is given by Eq. (6.1):

$$A_{OL\_SEAA} = A_{OL\_OPAM} \cdot \frac{R_{heat}}{R_{DEG}} \tag{6.1}$$

The performance of the single-ended analog proportional temperature controller in the tracking mode is shown in Fig. 6.4. The measurement was done at room temperature, and the control voltage of microhotplate 1 was increased in steps of 100 mV. For example, a control voltage of 1.60 V produced a microhotplate temperature of approximately 355 °C. Microhotplate 2 was kept at a constant temperature of 200 °C and microhotplate 3 was kept at 350 °C.

The tracking nonlinearity at high temperatures comes from the second-order coefficient of the polysilicon temperature sensor.

The performance of the single-ended analog proportional temperature controller in the stabilization mode is shown in Fig. 6.5. The ambient temperature was ramped from −40 to 120 °C in steps of 5 °C. A control voltage of 1.53 V was applied, which produced a microhotplate temperature of 331 °C. The steady-state error of the proportional temperature controller over the operating temperature range is less than 1% of the preset microhotplate temperature.

The discrepancy between ambient temperature and the bulk-silicon chip temperature as a function of the microhotplate temperature in a DIL ceramic package is shown in Fig. 6.6. The measurement was done at room temperature, and the control voltage was increased in steps of 100 mV heating the membrane from room temperature to 355 °C. The maximum temperature discrepancy due to the heating of the microhotplate was less than 3.5 °C.

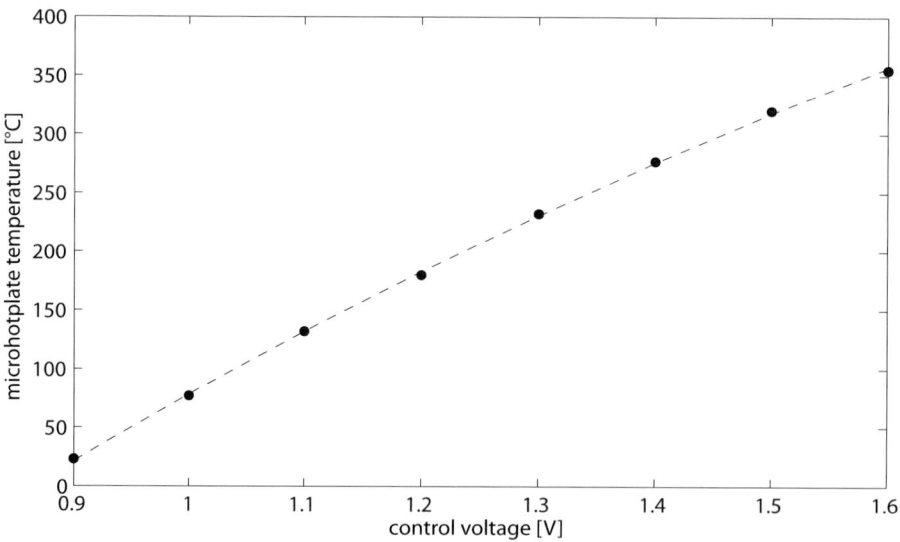

Fig. 6.4. Performance of the single-ended analog proportional temperature controller in the tracking mode

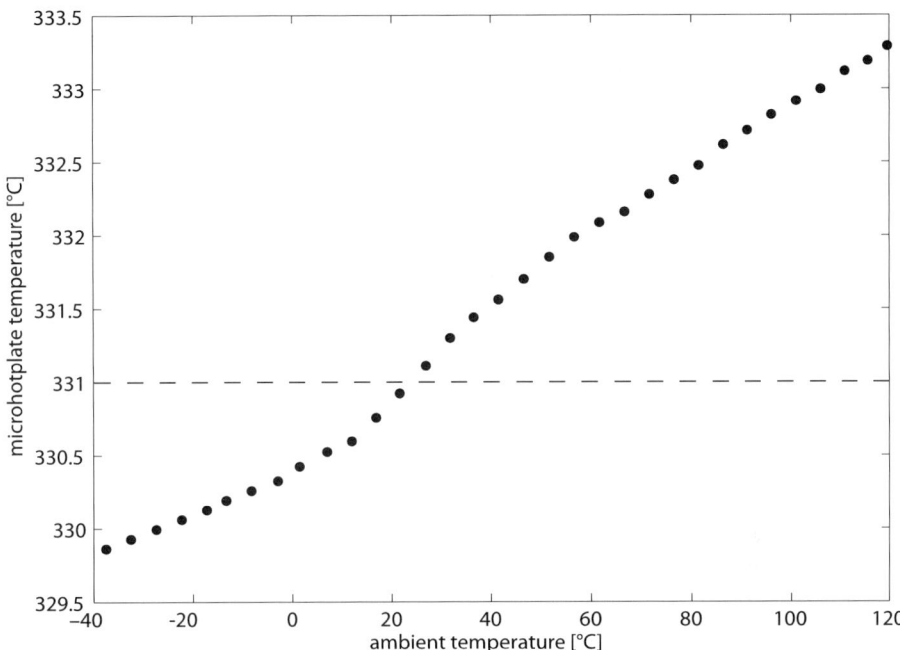

Fig. 6.5. Performance of the single-ended analog proportional temperature controller in the stabilization mode

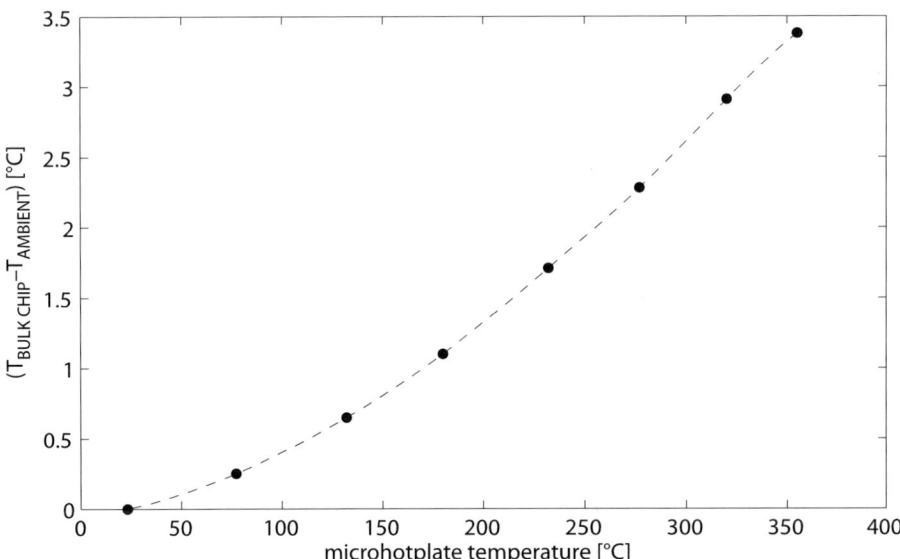

Fig. 6.6. Discrepancy between ambient temperature and bulk-chip temperature as a function of the microhotplate temperature for the single-ended analog architecture (DIL package)

6.2 Differential Mixed-Signal Architecture

6.2.1 System Description

A simplified block diagram of the differential mixed-signal architecture is shown in
Fig. 6.7 [132,146]. The differential mixed-signal architecture comprises a temperature
sensor on the bulk chip as described in Sect. 5.1.2, three digital proportional-integral-
derivative (PID) microhotplate-temperature controllers as described in Sect. 5.1.4,
three logarithmic converters for the readout of the sample resistances as described in
Sect. 5.1.3, three microhotplates with small dimensions equipped with a polysilicon
temperature sensor as described in Sect. 4.1, five pseudo-differential 10-bit successive-
approximation analog-to-digital converters (ADCs) for reading out three microhotplate temperatures, the bulk chip temperature, and the multiplexed sensitive material
(sample) resistances, five programmable offset compensators, three flash digital-to-
analog converters (DACs), a bias generator for the reference voltages of the ADCs
and the DACs, three pairs of dedicated serial lines for testing with an FPGA, and the
I^2C serial interface to handle the communication with external units. The differential
mixed-signal microsystem has been processed in 0.8 μm CMOS technology, which
resulted in a die size of 9.45×5.0 mm^2. A micrograph of the microsystem is shown in
Fig. 6.8.

Figure 6.9 shows a micrograph of the microhotplate with smaller dimensions than
that of the microhotplate used in the microsystems presented before. The membrane
is also 500 by 500 μm^2 in size, but the circular heated area (microhotplate) has a diameter of only 100 μm. The inner heater consists of a polysilicon ring with two heating arms connected in parallel. The connections between the polysilicon and the
aluminum-metallization lines are not placed on the heated area in order to reduce the
risk of electromigration. Moreover, due to the large distance between heated area and
polysilicon/Al-contacts the heat flow through the metal lines is massively decreased,
since there is an effective thermal decoupling. As a consequence, the power efficiency
of the microhotplate is increased, and its thermal resistance is 10.0 °C/mW.

6.2.2 Temperature Control Loop

The differential mixed-signal architecture has three temperature control loops (one
per microhotplate). Fig. 6.10 shows the schematic of one of them. The microhotplate
temperature is measured using a polysilicon resistor as temperature sensor (R_T). The
resistor is biased with a temperature-independent current source (I_{REF}). The voltage
drop across the resistor is low-pass filtered and converted to the digital domain using
a 10-bit pseudo-differential, successive-approximation ADC. The pseudo-differential
ADC is a slight modification of the successive-approximation ADC of the analog library of austria*micro*systems, the comparator of which has been replaced by a differential comparator, and to which a second DAC was added. The inputs of the digital
PID temperature controller include the feedback signal of the microhotplate temperature (T_M) and the 10-bit reference temperature (T_{REF}). The output of the digital temperature controller is converted to the analog domain using a 10-bit DAC ($V_{CONTROL}$).

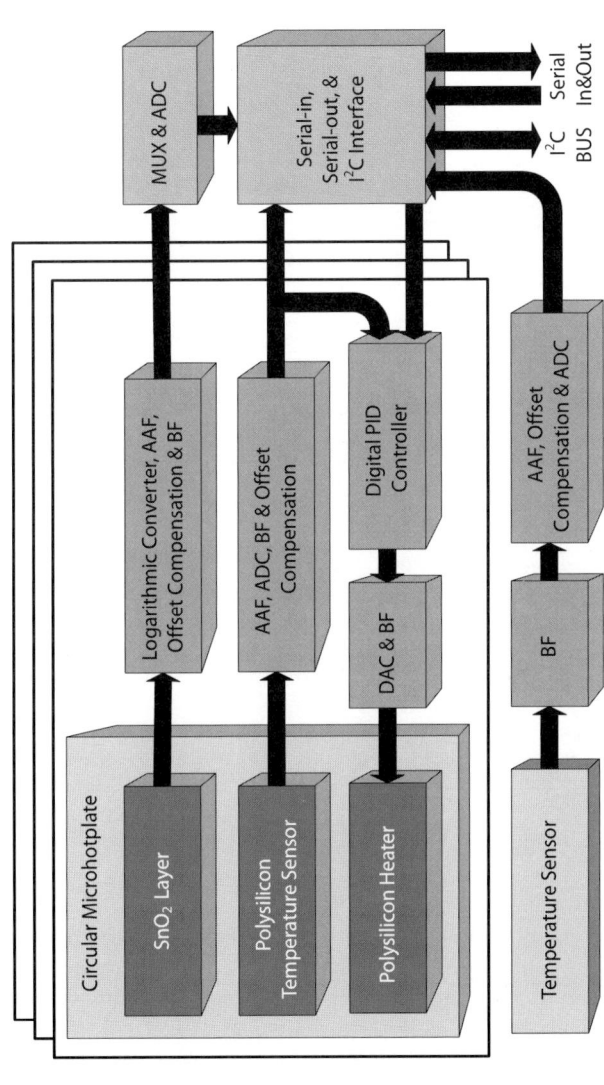

Fig. 6.7. Block diagram of the differential mixed-signal architecture. AAF: Anti-aliasing filter; BF: Buffer; MUX: Multiplexer

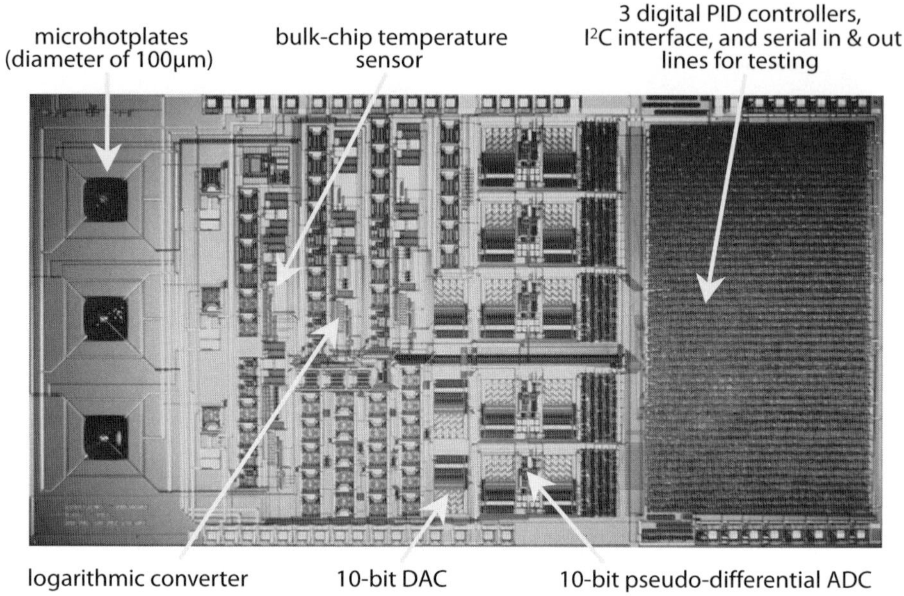

microhotplates
(diameter of 100μm)

bulk-chip temperature
sensor

3 digital PID controllers,
I²C interface, and serial in & out
lines for testing

logarithmic converter   10-bit DAC   10-bit pseudo-differential ADC

Fig. 6.8. Micrograph of the differential mixed-signal hotplate-based microsystem

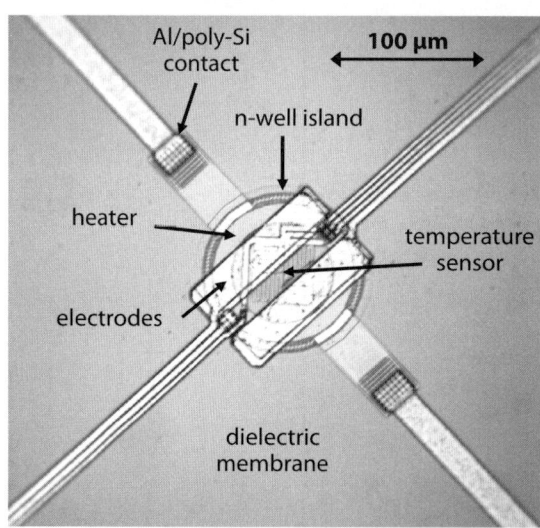

Al/poly-Si
contact

100 μm

n-well island

heater

temperature
sensor

electrodes

dielectric
membrane

Fig. 6.9. Micrograph of the small
microhotplate with a polysilicon
temperature sensor

$V_{CONTROL}$ is applied to a voltage-follower, since the DAC has a very high output impedance. The output of the voltage-follower drives the power transistor (M_1), which can be switched off for recalibration of the temperature sensor using the power-down transistor (M_2). The power transistor provides the current to the polysilicon heater (R_{heat}) and has a degeneration resistor (R_{DEG}) at the source, which limits the drain current. The differential architecture of this temperature control loop improves the power supply rejection ratio.

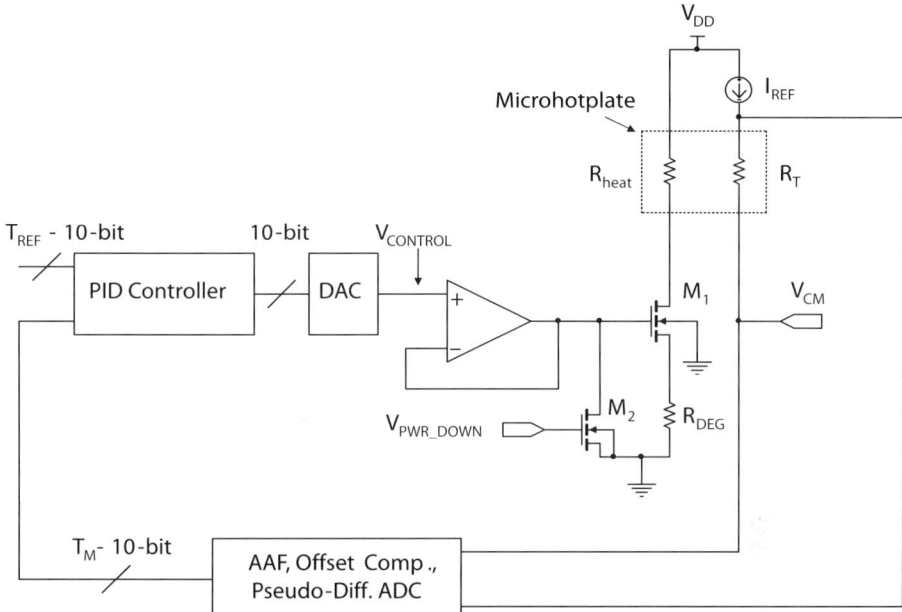

Fig. 6.10. Temperature control loop of the differential mixed-signal architecture

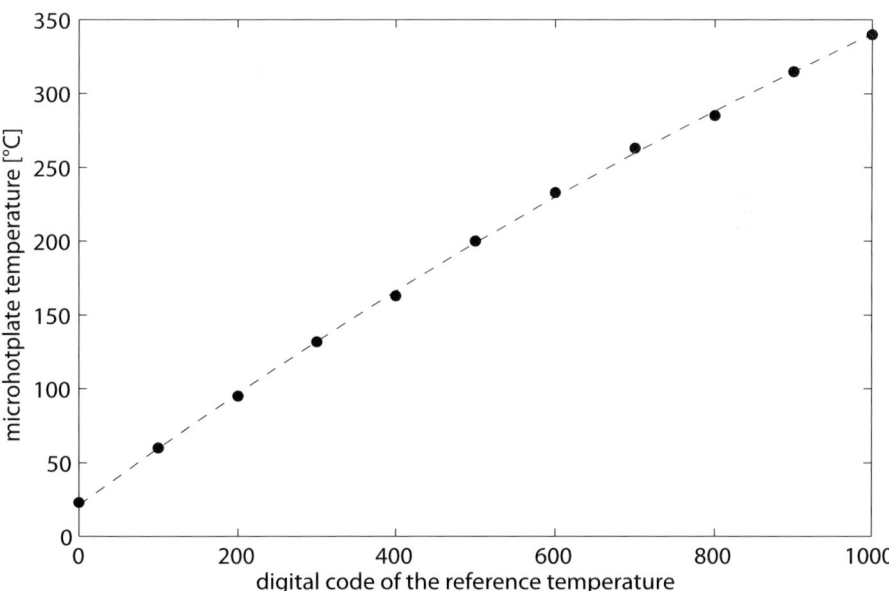

Fig. 6.11. Performance of the digital PID temperature controller of the differential mixed-signal architecture in the tracking mode

The tracking mode performance of the on-chip digital temperature controller of the differential mixed-signal architecture is shown in Fig. 6.11. The measurement was done at room temperature, and the digital code of the reference temperature of micro-hotplate 1 was increased in steps of 100 digits. Microhotplates 2 and 3 were switched off during the measurement. The resolution of the digital temperature controller was better than 2 °C.

6.2.3 I²C Serial Interface, Instruction Set, and Register Bank

The chip communication is handled by the inter-integrated circuit (I²C) serial interface. The microhotplate temperatures, the bulk chip temperature and the sensitive-material resistances are transferred to a PC using the I²C serial interface. The parameters of the three digital controllers, the reference temperatures, the dedicated logic for the serial-in/serial-out lines (used for chip testing and for development of new control algorithms), and the power-down mode are programmed using the I²C serial interface.

The I²C addressing scheme and the format of incoming messages (Format A and Format B) are the same as explained in Sect. 5.1.5.

The operation codes or the instruction set are defined as follows (Tab. 6.1):

Table 6.1. Operation codes of the differential mixed-signal architecture

Opcode	Binary Format	Description	Format
PWR UP	00000001	Enter full-power mode	A
PWR DN	00000010	Enter low-power mode	A
READ	00000011	Read ADC values	A
STREAM	00000101	Stream (read many) ADC values	A
GET n	$01n_{5-0}$	Get the value of register n	A
SET $n\ x$	$10n_{5-0}x_{15-0}$	Set the value of register n to x	B

In this microsystem architecture, the I²C recipient is also hard-coded to the value 01011000. The I²C senders identifies both, the message format and the address of the chip.

I²C sender

C/$\overline{\text{D}}$	a_6	a_5	a_4	a_3	a_2	a_1	a_0

The C/$\overline{\text{D}}$ bit is 1 for a message of format C. The format C, which is used when the chip receives a GET command, is the same as explained in Sect. 5.1.5.

The format D is issued in response to the STREAM and READ operation codes, containing the ten-bit values C, T, M, S, for chip temperature, test signal (used for testing the ADC), three microhotplate temperatures, and three sensitive-material resistances. The total length is 13 bytes. The ADC for the measurement of the sensitive-material resistance is multiplexed between the three microhotplates and the test signal in order to save silicon real estate.

Format D

I²C recipient	I²C sender	C_{9-2}	$Test_{9-2}$	$M_{0,9-2}$	$S_{0,9-2}$	C_{1-0} $M_{0,1-0}$ $S_{0,1-0}$ $Test_{1-0}$
				$M_{1,9-2}$	$S_{1,9-2}$	C_{1-0} $M_{1,1-0}$ $S_{1,1-0}$ $Test_{1-0}$
				$M_{2,9-2}$	$S_{2,9-2}$	C_{1-0} $M_{2,1-0}$ $S_{2,1-0}$ $Test_{1-0}$

The PWR DN command activates the power-down mode. In this mode, all the analog blocks are switched-off thus reducing the power consumption from 38 mA (all blocks of the chip are active, one microhotplate at 300 °C) to 6 mA (only the digital blocks are active).

The register bank of the differential mixed-signal architecture is shown in Tab. 6.2.

Table 6.2. Register bank of the differential mixed-signal architecture

Word	Bits	Description
0–3	15–0	*multi-purpose bits:* analog switches for offset compensation and calibration, loop selection (on-chip controller or FPGA controller), etc.
4	15–0	delay between streamed samples (in clk cycles) of the I²C and dedicated serial lines
5	15–0	sampling period of the digital controllers
6–8	9–0	reference or set-point temperatures of the three microhotplates
9–26	10–0	PID coefficients of the three microhotplates: $(a_0, a_1, a_2, c, b_1, b_2)_{0-2}$

The 1 Mbit/s serial-in and serial-out lines connect the microhotplates to an FPGA. The FPGA can be loaded with the PID control algorithm for testing or with new control algorithms. After loading the data to the FPGA, multiplexers close the temperature control loops through the FPGA.

The serial-out is realized as three serial lines that deliver a transmission word at intervals specified with register 4. A transmission word has a length of 20 bit and consists of two concatenated 10-bit words. Each transmission word is preceded by a start bit. The length of a bit is configurable with 5 bits of register 0.

Depending on whether the on-chip PIDs are utilized or the FPGA external feedback loops are chosen, the semantics of the transmission word change.

Serial-out: on-chip feedback-loop (PID)

$T_{REF} - T_{\text{Microhotplate } 9...0}$	Controller Actuation (on-chip PID) $_{9...0}$

Serial-out: FPGA feedback-loop

Microhotplate Temperature $_{9...0}$	– – –

The "serial-in" is only used in the FPGA external feedback-loop mode. The transmission word length is 10 bit. The speed is the same as that of the sending port. Also, each transmission operation is preceded by a start bit.

Serial-in: FPGA feedback-loop

Controller Actuation (from the FPGA) $_{9...0}$

6.3 Digital Array Architecture

6.3.1 System Description

As discussed in Sect. 4.4, microhotplates with integrated MOS-transistor-heaters offer the advantage of a relatively low power consumption. A further reduction of the power consumption can be achieved by implementing a fully digital temperature control. Such a digital architecture improves the robustness of the control and read-out scheme and entails a smaller chip size, especially if the design can be fabricated in a CMOS process with small minimum gate length.

The block diagram in Fig. 6.12 depicts the system architecture, emphasizing its strong modularity [147–149]. The voltage drop across the temperature sensor is converted by an analog-to-digital (A/D) converter into a digital signal. The A/D converter has been realized as a 10-bit successive-approximation A/D converter. An internal bandgap with a nominal voltage of 1.26 V or an external voltage can be used as reference voltage for the A/D converter.

The converted temperature signal serves as input to the digital PID temperature controller so that the temperature control loop is closed (see Sect. 6.3.3 for a detailed description of the temperature control loop).

The three independent controllers provide individual temperature regulation for each hotplate. The MOS heating transistors are driven in the pulse-density modulation mode by a first-order $\Sigma\Delta$-modulator. The A/D converter and a multiplier are the circuitry parts that consume the largest fractions of the chip-area. For this reason, they have been realized only once and are accessed in a time-sharing mode. The control unit takes care of scheduling the different operations and of controlling the access of all hotplates to the shared multiplier and analog-to-digital converter.

For each hotplate, an individual bias current is generated and applied to the metal-oxide-covered electrodes. The voltage drop across the resistive material is then converted to a digital value with the same A/D converter that is also used for the temperature control loop. The bias currents can be individually adjusted with a 10-bit value in a range from 0.1 μA to 90 μA. Therefore, the resistance of the SnO_2 material that can vary over orders of magnitude (from a few 100 Ω to 10 MΩ, depending on the addition of noble-metal catalysts), can be measured with a sufficient accuracy of about ±0.05% (minimum 10 Ω) of the actual resistance value. All parameters, such as target temperature, bias currents, PID coefficients and some control flags are stored within the control unit and are accessed via an I^2C interface as it was described in previous sections.

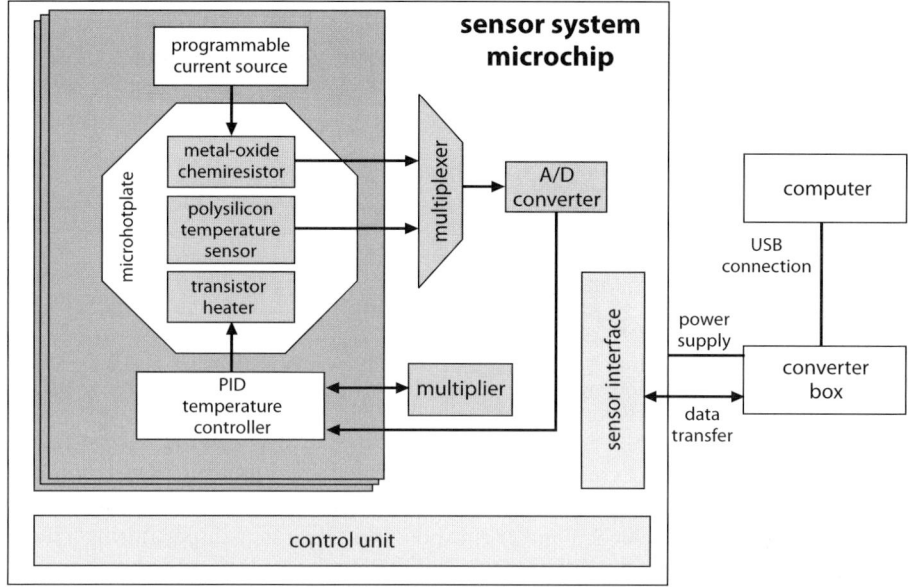

Fig. 6.12. Block diagram of the digital sensor system architecture

The chip is a standalone microsensor system that does not need any external measurement equipment for sensor control and readout. The sensor system chip has been connected to a computer via an I^2C-to-USB converter box, i.e., in this box is a microcontroller that translates the I^2C format coming from the chip into USB format for the computer or laptop. The power supply of the chip is also provided by the USB connection. The sensor system can be read out directly by a microcontroller and is, therefore, well suited for handheld devices or distributed sensor networks.

6.3.2 Microhotplate and Chip Layout

A cross-sectional schematic and layout of a microhotplate with MOS-transistor heater was already shown and discussed in Sect. 4.4. A slightly different layout was chosen for the monolithic system integration. The idea of a symmetric microhotplate layout was preserved (Fig. 6.13). The contact to the transistor gate is located in a corner of the octagonal PMOS ring transistor, which consists of 8 identical transistor segments and features a gate length of 5 µm and an overall gate width of 720 µm. The source and drain contacts were split so that four metallic leads of equal width provide the electrical connections to the transistor.

Two identical polysilicon temperature sensors with a nominal resistance value of 10 kΩ are located in the membrane center. One resistor is connected to the temperature controller, the other sensor is totally decoupled from the circuitry. This second temperature sensor can be directly accessed via bond pads in a four-point configuration. It enables an accurate calibration and a verification of the temperature controller

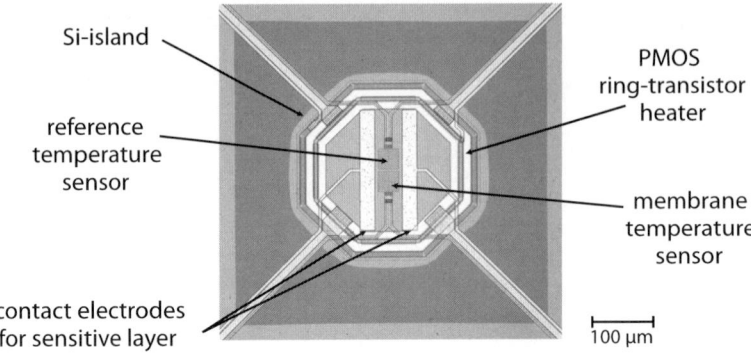

Fig. 6.13. Chip micrograph showing the membrane and the integrated PMOS transistor heater

performance. The dynamic characteristics of the controller can be monitored without interfering with the circuitry. The chip has been fabricated according to the process sequence described in Sect. 4.1.2. Figure 6.14 shows the fabricated chip featuring a floor plan with three distinct sections that include the digital circuitry at the bottom with the temperature controller and the interface, the analog circuitry in the center and the three micro-hotplates at the top.

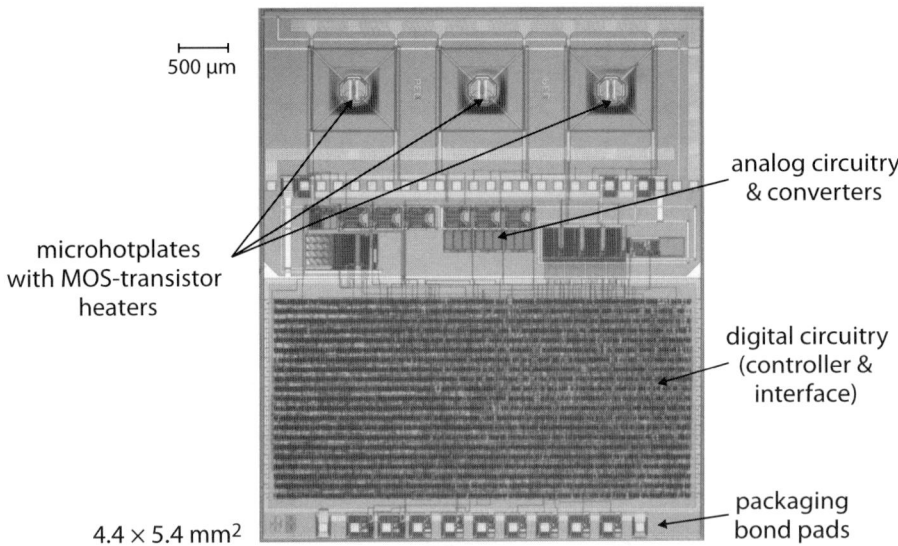

Fig. 6.14. Micrograph of the chip with microhotplate array and circuitry

The packaging strategy presented in Sect. 5.1.6 can be directly applied to this array chip (see Fig. 6.15). The number of necessary bond pads is only five, since this chip takes full advantage of the CMOS-MEMS approach. The number of bondpads would increase in proportion to the number of sensors if an integrated interface had not been used. The three different coatings of the chip are depicted in Fig. 6.16. The first coating is a 0.2 wt% Pd-doped nanocrystalline SnO_2 (a), which was used for most of the chemical measurements throughout this book. As mentioned in Sect. 2.3 describing the metal-oxide sensor basics, this material exhibits a high sensitivity towards CO. The second material is pure SnO_2, which is optimal for NO_2 sensing and features a slightly different film morphology (b). The third material with 3.0 wt% Pd appears more greyish (c). This material is more sensitive to hydrocarbons such as CH_4. The darker color is a consequence of the higher palladium content.

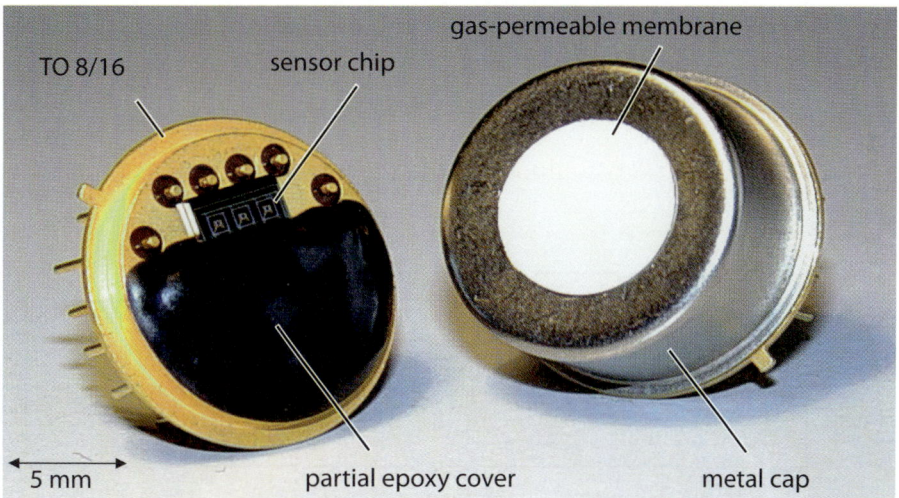

Fig. 6.15. Photo of a packaged sensor chip. The partial epoxy cover enables free analyte access to the chemical sensor area. The metal cap with the gas-permeable membrane provides mechanical and dust protection

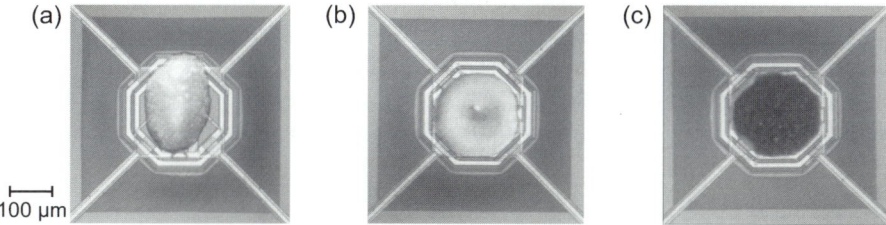

Fig. 6.16. Micrograph of the three different sensing materials: (a) SnO_2 with 0.2 wt% Pd, (b) undoped SnO_2, (c) SnO_2 with 3 wt% Pd

6.3.3 Temperature Control Loop and Sensor Resistance Readout

In Figure 6.17, the schematic of the temperature control loop is shown. The temperature on the microhotplate is measured using a poly-silicon resistor as a temperature sensor (R_{temp}), with a nominal resistance of approximately 10 kΩ. The resistor is biased with an adjustable, temperature-independent current source (I_{temp}) to cope with process variations. The voltage drop across the resistor is converted to a 10-bit digital value. This value is fed back to the digital PID temperature controller, which is implemented as a recursive filter with a biquadratic transfer function. The reference temperature is given as a 10-bit value, the three PID coefficients are 11-bit signed values that can be set individually for each PID. The actuation value calculated by the PID is then converted to a digital bit-stream using a digital first-order ΣΔ modulator with a sampling rate of half the clock frequency.

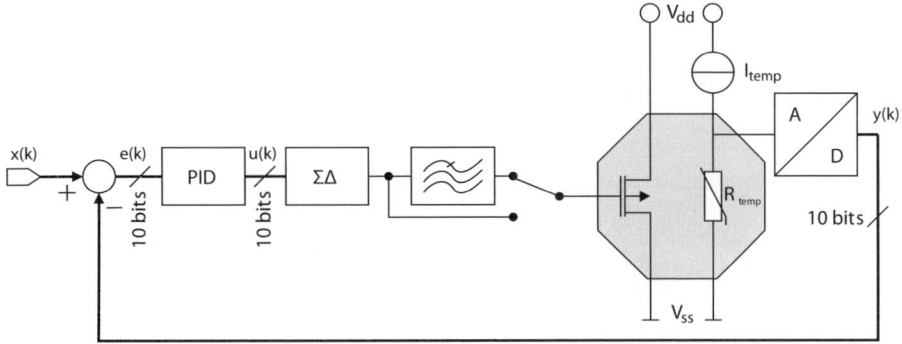

Fig. 6.17. Schematic of the temperature control loop

The MOSFET heater is directly driven with this bit stream or optionally, the bit stream can be routed through a low pass filter. Early experiments revealed that driving the heater transistor with the raw bit stream might cause some electromigration due to the large peak currents that occur. The low-pass filter was therefore included to avoid such current peaks in order to enhance the device stability and life time.

As the thermal capacitance and resistance of the hotplate provide a thermal low-pass transfer function (with the dominant pole corresponding to a characteristic time of 10–20 ms, depending on the fabrication process), the ΣΔ modulator driving the hotplate constitutes a linear noise-shaping DAC with an output in the thermal domain.

6.3.4 Circuitry Assessment

The sensor can either be mounted in a DIL28 package (dual-inline package with 28 pins) for testing and calibration purposes or in a TO-package (Fig. 6.15), which is then directly mounted on the I^2C-to-USB converter-box (Sect. 6.3.1). The fully packaged chip is connected via the USB-to-I^2C connector box to a laptop or palm top computer. A LABVIEW™ program was implemented as user-friendly interface for setting the sensor system parameters, for generating the individual temperature profiles for the hotplates, and to display the measurement results. Any arbitrary waveform can be generated to modulate the temperature on the hotplates (dynamic or static operation). The software translates the digital sensor system output into the corresponding microhotplate temperatures and sensor resistance values. The sampling rate is seven complete data sets per second.

The performance of the temperature controller was measured in the tracking mode. Figure 6.18 shows a graph, where the temperature of one of the three microhotplates is kept at a constant temperature of 300 °C, the temperature of the second microhotplate is modulated using a sine wave of 10 mHz, while rectangular temperature steps of 150 °C, 200 °C, 250 °C, 300 °C, and 350 °C have been applied to the third microhotplate. Temperature measurements on one of the hotplate that has been operated at constant temperature in the stabilization mode showed a variation of less than 1 °C, even though the temperature of the neighboring hotplates was, at the same time, modulated dynamically (sine wave, ramp, steps). This is a consequence of the individual hotplate temperature control, without which thermal crosstalk between the hotplates would have been clearly detectable. The power dissipation of the chip is approximately 190 mW, when all three hotplates are simultaneously heated to 350 °C. In the power-down mode, the power consumption is reduced to 8.5 mW.

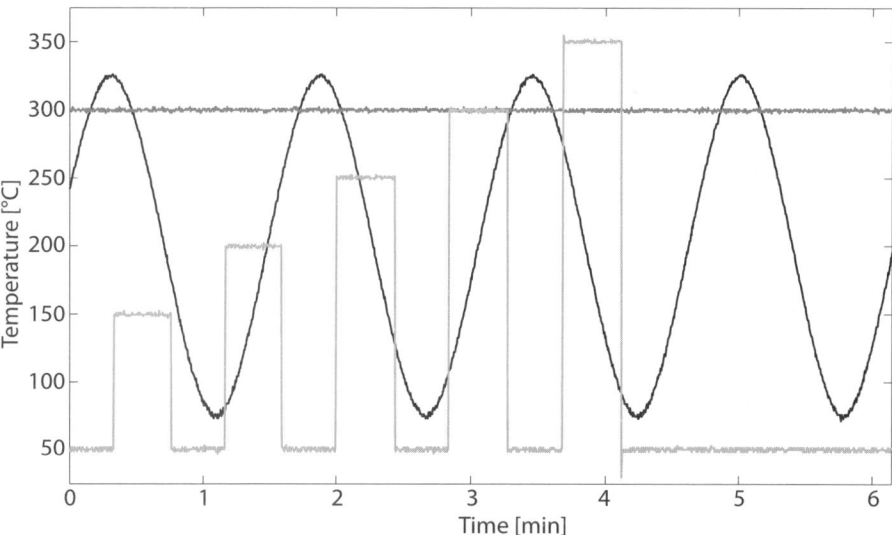

Fig. 6.18. Temperature tracking in the PID control mode: one hotplate is kept at constant temperature, the temperature of the second microhotplate is modulated in a sinusoidal way. Rectangular temperature steps are applied to the third microhotplate

6.3.5 Gas Test Measurements

First chemical test measurements have been conducted with the array chip. Figure 6.19 shows the results that have been obtained simultaneously from three microhotplates coated with different tin-dioxide-based materials at operation temperatures of 280 °C and 330 °C in humidified air (40% relative humidity at 22 °C). The first microhotplate (μHP1) is covered with a Pd-doped SnO_2 layer (0.2 wt% Pd), which is optimized for CO-detection, whereas the sensitive layer on microhotplate 3 contains 3 wt% Pd, which renders this material more responsive to CH_4. The material on microhotplate 2 is pure tin oxide, which is known to be sensitive to NO_2. Therefore, the electrodes on microhotplate 2 do not measure any significant response upon exposure to CO or methane. The digital register values can be converted to resistance values by taking into account the resistor bias currents [147, 148]. The calculated baseline resistance of microhotplate 1 is approximately 47 kΩ, that of hotplate 2 is 370 kΩ and the material on hotplate 3 features a rather large resistance of nearly 1 MΩ.

As it is evident from Fig. 6.19, the responses of the three sensitive materials to the test gases CO and CH_4 are very different. The lightly Pd-doped (0.2%) tin dioxide shows large responses to CO, and very small responses to CH_4, whereas the heavily Pd-doped (3%) tin oxide exhibits comparably smaller responses to CO, but also

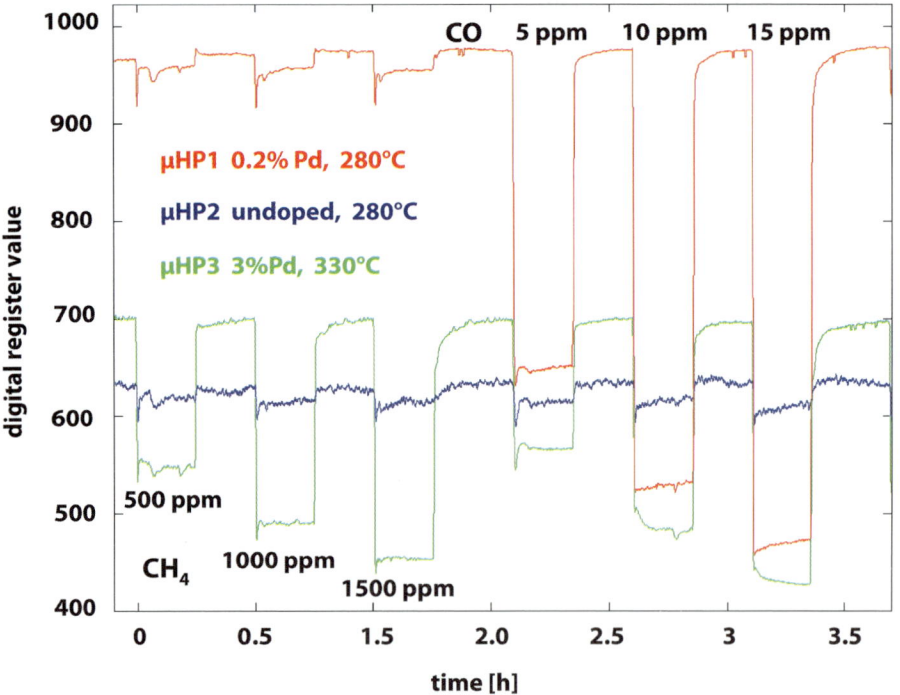

Fig. 6.19. Sensor responses (raw digital values) of three microhotplates with different sensitive layers upon exposure to CO and CH_4 (40% r.h.)

responds to hydrocarbons, as in this case, methane. Excellent discrimination of the two gases was achieved with detection limits of less than 1 ppm for CO, and less than 100 ppm for CH_4. The pure tin dioxide shows hardly any response to CO or methane but has been found to provide signals upon exposure to nitrogen dioxide at concentrations as low as 10 ppb [150].

7

Conclusion and Outlook

The central topic of the book was the integration of microhotplate-based metal-oxide gas sensors with the associated circuitry to arrive at single-chip systems. Innovative microhotplate designs, dedicated post-CMOS micromachining steps, and novel system architectures have been developed to reach this goal. The book includes a multitude of building blocks for an application-specific sensor system design based on a modular approach.

Modelling and Simulations

A thermal model of the CMOS-based microhotplates was introduced that enables a prediction of the thermal characteristics for a given layout (Chap. 3). The importance of this model can be deduced from the fact that it enables a comprehensive simulation of the overall sensor system, which includes all necessary steps from the microhotplate layout to the final AHDL model of the sensor. Nonlinear effects as a consequence of the temperature dependence of the thermal conductivities of the different materials were also included in the FEM simulations. A main issue was the avoidance of fitting parameters. Despite such rigid boundary conditions an accordance between simulated and measured results within 5% for the thermal resistance of the microhotplates and within approx. 10% for the thermal time constants has been obtained. This is sufficient for system simulations – the fluctuations in the electronics characteristics are on the same order of magnitude. The observed discrepancy between measured and simulated thermal resistances was mainly attributed to the uncertainty in the thermal conductivities of the CMOS materials at higher temperatures. All values had been extracted for temperatures up to 125 °C, which is much lower than the common operating temperature of metal oxide sensors that is in the range of 200–400 °C. A more accurate thermal simulation would, therefore, require measurements of the thermal properties of CMOS thin-film materials at higher temperatures. Additional effects such as radiation-induced effects would have to be considered as well. Another question is the appropriate inclusion of the nanocrystalline tin-oxide droplet in the model. At the moment, reliable heat conductivity values of this porous material are not available. Nevertheless, a rough estimation of the droplet volume already led to good results for estimating the time constant of a coated microhotplate.

Microhotplate Design

The first device was a circular-shape microhotplate, which essentially consisted of CMOS-process materials (Sect. 4.1). The fabrication of this microhotplate required a minimum of post-CMOS processing steps. The electrochemical etching process used for the membrane release and the formation of the circular-shape Si island was optimized and can now be routinely applied on wafer-level.

In order to establish good electrical contact to the sensitive layer, it was necessary to coat the electrodes with a metal stack of Ti/W (diffusion barrier and adhesion layer) and Pt. The usage of a shadow mask during the metal deposition ensures full compatibility with other MEMS processing steps so that it is possible to fabricate various CMOS-MEMS devices on the same wafer.

The full CMOS compatibility of the annealing process of the sensitive-material that has been conducted at a temperature of 400 °C was demonstrated, so that post-processing steps can be performed on wafer level. An important issue concerns the necessity of the silicon island underneath the heated area. Based on the experimental results, it was shown that a thin polysilicon plate in the membrane center could replace the thick n-well silicon island. Such a plate suffices to achieve good temperature homogeneity across the microhotplate. The use of a purely dielectric membrane (including polysilicon plate) will allow for a substitution of the technologically expensive ECE-process by a regular KOH-wet etching step with an etch stop on the field oxide of the dielectric membrane. Moreover, dry-etching methods such as DRIE can be used, which will allow for a higher packing density of the micromachined structures.

The thermal resistance of the circular hotplate was measured to be 5.8 °C/mW for the coated and uncoated transducer. An increased thermal resistance is desirable for sensor arrays with one approach being the reduction of the heated membrane area. At the moment the smallest possible diameter of drop-deposited tin-oxide is 100 μm. A microhotplate with a heated area of 100 μm in diameter was fabricated and featured an increased thermal resistance of approximately 10 °C/mW.

The main goal of another microhotplate design was the replacement of all CMOS-metal elements within the heated area by materials featuring a better temperature stability. This was accomplished by introducing a novel polysilicon heater layout and a Pt temperature sensor (Sect. 4.3). The Pt-elements had to be passivated for protection and electrical insulation, so that a local deposition of a silicon-nitride passivation through a mask was performed. This silicon-nitride layer also can be varied in its thickness and with regard to its stress characteristics (compressive or tensile). This hotplate allowed for reaching operation temperatures up to 500 °C and it showed a thermal resistance of 7.6 °C/mW.

The third microhotplate design included a MOS-transistor heater embedded in a silicon island (Sect. 4.4). One advantage of this configuration is the reduction of the overall system power consumption, since no additional on-chip power transistor is needed to control the heating current. Operating temperatures up to 350 °C have been explored with this device. An almost linear relationship between microhotplate temperature and the source-gate voltage was observed for this transistor-heated hotplate, which renders the device suitable for any arbitrary temperature modulation protocol. An analytical transistor model coupled with the thermal microhotplate characteristics was developed to support the measurement results. The model has been based on established transistor models and the agreement between measured and simulated results was within 5–10% for temperatures up to 300 °C without the need to introduce additional fitting parameters.

Gas test measurements were performed with this device, and a novel sensor operation mode was developed and successfully demonstrated (Sect. 4.5): calorimetric-type signals could be recorded as changes in the source-gate voltage of the transistor while maintaining a constant microhotplate temperature.

There are still open issues with the CMOS-based microhotplates, such as the long-term stability of the microhotplate, which is of paramount importance for commercialization. For operating temperatures of less than 300 °C the temperature sensor drifts are acceptable. The quantitative assessment of drift effects of temperature sensors is, however, difficult and requires long-term measurements with a large number of devices. Moreover, the drifting behavior depends on the thermal history of the materials and, in most cases, hysteresis effects occur. Furthermore, the delamination of the nanocrystalline material from the micromachined substrate can also cause problems. Membrane buckling upon cyclic heating might lead to cracks in the sensitive layer. To assess long-term deterioration effects the devices have to be subjected to accelerated life-time measurements. Though there was no systematic investigation of longterm or aging effects of the devices, no major problems or system breakdowns were observed during extended sensor system operations of several weeks or mounths.

Integrated Monolithic Systems

A new class of microsensor systems comprising the microhotplates and the corresponding read-out and control circuitry were devised. Various system architectures of different complexity have been realized.

In a first device (Sect. 5.1), the temperature of a circular microhotplate has been regulated by a means of a digital controller. A logarithmic converter was introduced to cope with the wide resistance range of the tin oxide (wide range of initial material resistance and large range of gas-induced resistance changes), and to provide a first-order linearization of the sensor signal. An additional temperature sensor was used to measure the temperature of the bulk chip. The integrated I^2C interface manages the data transfer from the chip to a computer via a microcontroller.

The on-chip hotplate temperature controller had a resolution of better than ±2 °C. A high CO sensitivity was measured with a 0.2 wt % doped Pd nanocrystalline tin oxide, the resolution was 0.2 ppm CO, and the detection limit was 0.1 ppm CO. Such low concentrations are relevant for CO monitoring in environmental, industrial and household applications.

A transistor-outline (TO)-based prototype package was presented that showed favorable thermal characteristics. The bulk chip temperature increase owing to the hotplate and circuitry power dissipation was less than 1% of the selected microhotplate operation temperature (e.g., 2 °C for 300 °C hotplate temperature), which does not compromise the circuitry functionality. The package can hence be realized as a low-cost solution for commercialization, such as a plastic package with hotplate openings.

The features of the monolithic integrated sensor systems have not yet been fully exploited. The almost linear relationship between input reference voltage and microhotplate temperature renders the systems suitable for applying any temperature modulation protocol. Due their compatibility with other CMOS-based chemical sensors the microhotplates can be also combined with, e.g., polymer-based mass sensitive, calorimetric or capacitive sensors. The co-integration with such sensors can help to alleviate problems resulting from cross-sensitivities of tin-oxide based sensors to, e.g., volatile compounds such as hydrocarbons. A well-known problem is the cross-sensitivity of tin oxide to humidity or ethanol. The co-integration of a capacitive sensor, which does not show any sensitivity to CO, could help to independently assess humidity changes.

To overcome the temperature limits of CMOS integrated systems that are imposed by, e.g., the degradation of the CMOS metallization, a microhotplate with Pt-temperature sensor was also monolithically integrated with circuitry so that the hotplate operating temperature range could be extended to 500 °C (Sect. 5.2). The read-out of the comparatively low Pt temperature sensor resistance required the integration of a fully differential amplifier architecture.

Sensor-Array Systems

Three different sensor-array systems have been presented in this book. The first array features three circular microhotplates and an analog single-ended system architecture (Sect. 6.1). The readout and temperature control circuitry is repeated three times. The required chip real estate needed for the analog circuitry has been minimized. However, since no communication interface has been implemented, the number of bondpads that are needed to operate the system is still high. Hotplate temperature modulation can be performed through modulating the input reference voltages for the temperature controllers. The temperature variations on the hotplates then will reproduce the input signal waveform.

The second hotplate array (Sect. 6.2) is an extended version of the single-ended mixed-signal architecture described above. The temperature sensor read-out was changed to a differential configuration. The heated area of the microhotplates was reduced to a diameter of 100 μm. Three such microhotplates and the corresponding individual sensor read-out and temperature control circuits have been integrated with the sensor array, which also featured an I^2C-interface to connect off-chip to a PC. This interface can be used for data transfer and for setting the digital controller parameters.

The last and most advanced system presented in this book includes an array of three MOS-transistor-heated microhotplates (Sect. 6.3). The system relies almost exclusively on digital electronics, which entailed a significant reduction of the overall power consumption. The integrated I^2C interface reduces the number of required wire bond connections to only ten, which allows to realize a low-prize and reliable packaging solution. The temperature controllers that were operated in the pulse-density mode showed a temperature resolution of $\pm 1\,°C$. An excellent thermal decoupling of each of the microhotplates from the rest of the array was demonstrated, and individual temperature modulation on the microhotplates was performed. The three microhotplates were coated with three different metal-oxide materials and characterized upon exposure to various concentrations of CO and CH_4.

Applications and Outlook

More and more microhotplate-based chemical sensors become commercially available, which can be taken as a measure of success in the miniaturization efforts for this type of chemical sensor. The microsensor systems presented here will most probably represent the next generation of these sensors. The number of possible applications in industrial, environmental and household settings is large. Examples include the integration of sensor chips in mobile systems and distributed networks to monitor concentrations of environmentally relevant gases, or the surveillance of air quality to, e.g., avoid CO-poisoning in low energy houses. In particular the systems, which are capable of simultaneously monitoring three or more target gases, potentially serve a wide variety of applications such as gas hazard and leakage detection in household and industrial settings, indoor air quality monitoring in household and automobile applications, and they can be used as personal safety devices. The challenges on the way to commercial products include, e.g., to eliminate the need for frequent sensor recalibration and to increase the sensor stability and life time.

The presented integrated microdevices are not only compact sensor systems, but also constitute excellent research tools for sensor science. The influence of electrode materials and configurations can be studied, and optimal operating temperatures and the respective temperature modulation protocols can be established. The chemoresistive signals can be combined with simultaneously recorded calorimetric and, in future designs, thermoelectric signals. All recorded output signals can be processed by using data analysis and pattern recognition methods. An on-chip integration of a first-order data processing unit in the near future seems to be advantageous.

The microsystems may also serve potential applications in material science and in the growing field of nanotechnology. Microhotplates can be used for material processing, and, at the same time, for the monitoring of material properties such as the electrical resistance [10]. Moreover, the microsystems can be applied to determine thermal properties of new materials such as the melting point, especially when only small quantities of material are available [145], so that monolithic microhotplate-based devices are not only powerful sensor systems for a broad range of applications, but also new research tools for sensor science and nanotechnology.

References

1. G. Heiland and D. Kohl. "Physical and Chemical Aspects of Oxidic Semiconductor Gas Sensors", *Chemical Sensor Technology*, T. Seiyama, (Ed.), Elsevier, Amsterdam, Netherlands (1988).
2. M.J. Madou and S.R. Morrison. *Chemical Sensing with Solid State Devices*, Academic Press, New York, USA (1989).
3. W. Göpel and G. Reinhardt. "Metal Oxide Sensors: New Devices Through Tailoring Interface on the Atomic Scale", *Sensors Update, Vol. 1*, H. Baltes, W. Göpel, J. Hesse, Eds., Verlagsgesellschaft mbH: Weinheim, Germany (1991).
4. G. Sberveglieri. "Classical and novel techniques for the preparation of SnO$_2$ thin-film gas sensors", *Sensors and Actuators* **B6** (1992), 239–247.
5. N. Bârsan, M. Schweizer-Berberich, and W. Göpel. "Fundamental and practical aspects in the design of nanoscaled SnO$_2$ gas sensors: a status report", *Fresenius' Journal of Analytical Chemistry* **365** (1999), 287–304.
6. N. Bârsan, J.R. Stetter, M. Findlay, and W. Göpel. "High performance gas sensing of CO: Comparative tests for semiconducting (SnO$_2$-based) and for amperometric gas sensors", *Analytical Chemistry* **71** (1999), 2512–2517.
7. G. Sberveglieri, W. Hellmich, and G. Müller. "Silicon hotplates for metal oxide gas sensor elements", *Microsystem Technologies* **3** (1997), 183–190.
8. J.W. Gardner, V.K. Varadan, and O.O. Awadelkin. *Microsensors, MEMS and Smart Devices*, Wiley, New York, USA (2001).
9. I. Simon, N. Bârsan, M. Bauer, and U. Weimar. "Micromachined metal oxide gas sensors: opportunities to improve sensor performance", *Sensors and Actuators* **B73** (2001), 1–26.
10. B. Panchapakesan, D.L. DeVoe, M.R. Widmaier, R. Cavicchi, and S. Semancik. "Nanoparticle engineering and control of tin oxide microstructures for chemical microsensor applications", *Nanotechnology* **12** (2001), 336–349.
11. E. Comini, G. Faglia, G. Sberveglieri, Z.W. Pan, and Z.L. Wang. "Stable and highly sensitive gas sensors based on semiconducting oxide nanobelts", *Applied Physics Letters* **81** (2002), 1869–1871.
12. J.V. Hatfield, A. Armitage, S.A. Bell, and P.I. Neaves. "ASICs for integrated sensors", *Digest IEE Colloquium on Advances in Sensors*, London, UK (1995), 4/1–8.
13. G. Müller, P.P. Deimel, W. Hellmich, and C. Wagner. "Sensor fabrication using thin film-on-silicon approaches", *Thin Solid Films* **296** (1997), 157–163.
14. A. Hierlemann and H. Baltes. "CMOS-based chemical microsensors" *Analyst* **128** (2003), 15–28.

15. S. Semancik, R.E. Cavicchi, M.C. Wheeler, J.E. Tiffany, G.E. Poirier, R.M. Walton, J.S. Suehle, B. Panchapakesan, and D.L. DeVoe. "Microhotplate platforms for chemical sensor research", *Sensors and Actuators* **B77** (2001), 579–591.
16. N. Najafi, K.D. Wise, R. Mechant, and J.W. Schwank. "An integrated multi-element ultra-thin-film gas analyzer", *Digest IEEE Solid State Sensor and Actuator Workshop*, Hilton Head Island, SC, USA (1992), 19–22.
17. G.C. Cardinali, L. Dori, M. Fiorini, I. Sayago, G. Fagila, C. Perego, G. Sberveglieri, V. Liberali, F. Maloberti, and D. Tonietto. "A smart sensor system for carbon monoxide detection", *Analog Integrated Circuits and Signal Processing* **14** (1997), 275–296.
18. P.F. Ruedi, P. Heim, A. Mortara, E. Franzi, H. Oguey, and X. Arreguit. "Interface circuit for metal-oxide gas sensor", *Digest IEEE Custom Integrated Circuits Conference* (2001), 109–112.
19. Y.W. Mo, Y.Z. Okawa, K.J. Inoue, and K. Natukawa. "Low-voltage and low-power optimization of micro-heater and its on-chip drive circuitry for gas sensor array", *Sensors and Actuators* **A100** (2002), 94–101.
20. C. Hagleitner, A. Hierlemann, D. Lange, A. Kummer, N. Kerness, O. Brand, and H. Baltes. "Smart single-chip gas sensor microsystem" *Nature* **414** (2001), 293–296.
21. J.S. Suehle, R.E. Cavicchi, M. Gaitan, and S. Semancik. "Tin oxide gas sensor fabricated using CMOS micro-hotplates and in-situ processing", *IEEE Electron Device Letters* **14** (1993), 118–120.
22. J.A. Covington, F. Udrea, and J.W. Gardner. "Resistive gas sensor with integrated MOSFET micro hot-plate based on an analogue SOI CMOS process", *Proc. IEEE Conference on Sensors*, Orlando, FL, USA (2002), 1389–1394.
23. J. Wöllenstein, J.A. Plaza, C. Cané, Y. Min, H. Böttner, and H.L. Tuller. "A novel single chip thin film metal oxide array", *Sensors and Actuators* **B 93** (2003), 350–355.
24. S.A. Bota, A. Dieguez, J.L. Merino, R. Casanova, J. Samitier, and C. Cané. A Monolithic Interface Circuit for Gas Sensor Arrays: Control and Measurement", Analog Integrated Circuits and Signal Processing **40** (2004), 175–184.
25. R.P. Manginell, J.H. Smith, and A.J. Ricco. "An overview of micromachined platforms for thermal sensing and gas detection", *Proc. of the SPIE* **3046** (1997), 273–284.
26. P. Hille and H. Strack. "A heated membrane for a capacitive gas sensor", *Sensors and Actuators* **A32** (1992), 321–325.
27. R. Aigner, M. Dietl, R. Katterloher, and V. Klee. "Si-planar-pellistor: designs for temperature modulated operation", *Sensors and Actuators* **B33** (1996), 151–155.
28. P.P. Tsai, I.C. Chen, and C.J. Ho. "Ultralow power carbon monoxide microsensor by micromachining techniques", *Sensors and Actuators* **B76** (2001), 380–387.
29. C. Dücsö, E. Vázsonyi, M. Ádám, I. Szabó, I. Bársony, J.G.E. Gardeniers, and A. van den Berg. "Porous silicon bulk micromachining for thermally isolated membrane formation", *Sensors and Actuators* **A60** (1997), 235–239.
30. D. Briand, B. van der Schoot, N.F. de Rooij, H. Sundgren, and I. Lundstrøm. "A low-power micromachined MOSFET gas sensor", *Journal of Microelectromechanical Systems* **9** (2000), 303–308.
31. D. Briand, H. Sundgren, B. van der Schoot, I. Lundstrøm, and N.F. de Rooij. "Thermally isolated MOSFET for gas sending application", *IEEE Electron Device Letters* **22** (2001), 11–13.
32. I. Simon and M. Arndt. "Thermal and gas-sensing properties of a micromachined thermal conductivity sensor for the detection of hydrogen in automotive applications", *Sensors and Actuators* **A97–98** (2002), 104–108.
33. N. Najafi, K.D. Wise, and J.W. Schwank. "A Micromachined Ultra-Thin-Film Gas Detector", *IEEE Trans. Electron Devices* **41** (1994), 1770–1777.

34. C.L. Johnson, J.W. Schwank, and K.D. Wise. "Integrated ultra-thin-film gas sensors", *Sensors and Actuators* **B20** (1994), 55–62.
35. V. Demarne and A. Grisel. "An integrated low-power thin-film CO gas sensor on silicon", *Sensors and Actuators* **13** (1988), 301–313.
36. S.K.H. Fung, Z.N. Tang, P.C.H. Chan, J.K.O. Sin, and P.W. Cheung. "Thermal analysis and design of a micro-hotplate for integrated gas-sensor applications", *Sensors and Actuators* **A54** (1996), 482–487.
37. L.Y. Sheng, Z.N. Tang, J. Wu, P.C.H. Chan, and J.K. O. Sin. "A low-power CMOS compatible integrated gas sensor using maskless tin oxide sputtering", *Sensors and Actuators* **B49** (1998), 81–87.
38. A. Friedberger, P. Kreisl, E. Rose, G. Müller, G. Kuhner, J. Wöllenstein, and H. Böttner. "Micromechanical fabrication of robust low-power metal oxide gas sensors", *Sensors and Actuators* **B 93** (2003), 345–349.
39. F. Solzbacher, C. Imawan, H. Steffes, E. Obermeier, and M. Eickhoff. "A new SiC/HfB$_2$ based low power gas sensor", *Sensors and Actuators* **B77** (2001), 111–115.
40. U. Dibbern. "A Substrate for Thin-Film Gas Sensors in Microelectronic Technology", *Sensors and Actuators* **B2** (1990), 63–70.
41. W.Y. Chung, C.H. Shim, S.D. Choi, and D.D. Lee. "Tin Oxide Microsensor for LPG Monitoring", *Sensors and Actuators* **B20** (1994), 139–143.
42. J.W. Gardner, A. Pike, N.F. de Rooij, M. Koudelka-Hep, P.A. Clerc, A. Hierlemann, and W. Göpel. "Integrated Array Sensor for Detecting Organic-Solvents", *Sensors and Actuators* **B26** (1995), 135–139.
43. D.D. Lee, W.Y. Chung, M.S. Choi, and J.M. Baek. "Low-power micro gas sensor", *Sensors and Actuators* **B33** (1996), 147–150.
44. V. Guidi, G.C. Cardinali, L. Dori, G. Faglia, M. Ferroni, G. Martinelli, P. Nelli, and G. Sberveglieri. "Thin-film gas sensor implemented on a low-power-consumption micromachined silicon structure", *Sensors and Actuators* **B49** (1998), 88–92.
45. S. Astié, A.M. Gué, E. Scheid, L. Lescouzeres, and A. Cassagnes. "Optimization of an integrated SnO$_2$ gas sensor using a FEM simulator", *Sensors and Actuators* **A69** (1998), 205–211.
46. S. Astié, A.M. Gué, E. Scheid, and J.P. Guillemt. "Design of a low power SnO$_2$ gas sensor integrated on silicon oxynitride membrane" *Sensors and Actuators* **B67** (2000), 84–88.
47. A. Götz, I. Grácia, C. Cané, E. Lora-Tamayo, M.C. Horrillo, J. Getino, C. García, and J. Gutiérrez. "A micromachined solid state integrated gas sensor for the detection of aromatic hydrocarbons", *Sensors and Actuators* **B44** (1997), 483–487.
48. D. Briand, A. Krauss, B. van der Schoot, U. Weimar, N. Bârsan, W. Göpel, and N.F. de Rooij. "Design and fabrication of high-temperature micro-hotplates for drop-coated gas sensors", *Sensors and Actuators* **B68** (2000), 223–233.
49. P.C.H. Chan, T. Gui Zhen, S. Lie Yi, R.K. Sharma, T. Zhenan, J.K.O. Sin, I.M. Hsing, and W. Yangyuan. "An integrated gas sensor technology using surface micro-machining", *Sensors and Actuators* **B82** (2002), 277–283.
50. A. Splinter, O. Bartels, and W. Benecke. "Thick porous silicon formation using implanted mask technology", *Sensors and Actuators* **B76** (2001), 354–360.
51. A. Stoffel, A. Kovacs, W. Kronast, and B. Müller. "LPCVD against PECVD for micromechanical applications.", *Journal of Micromechanics and Microengineering* **6** (1996), 1–13.
52. F. Udrea, J.W. Gardner, D. Setiadi, J.A. Covington, T. Dogaru, C.C. Lua, and W.I. Milne. "Design and simulations of SOI-CMOS micro-hotplate gas sensors", *Sensors and Actuators* **B78** (2001), 180–190.

53. M. Graf, A. Gurlo, N. Bârsan, U. Weimar, and A. Hierlemann. "Microfabricated gas sensor systems with sensitive nanocrystalline metal-oxide films", Journal of Nanoparticle Research (2006), 8, 823–839.

54. R.E. Cavicchi, J.S. Suehle, K.G. Kreider, B.L. Shomaker, J.A. Small, M. Gaitan, and P. Chaparala. "Growth of SnO₂ Films on Micromachined Hotplates", *Applied Physics Letters* **66** (1995), 812–814.

55. N.O. Savage, S. Roberson, G. Gillen, M.J. Tarlov, and S. Semancik. "Thermolithographic patterning of sol-gel metal oxides on micro hot plate sensing arrays using organosilanes", *Analytical Chemistry* **75** (2003), 4360–4367.

56. V. Demarne and A. Grisel. "A New SnO₂ Low-Temperature Deposition Technique for Integrated Gas Sensors", *Sensors and Actuators* **B15** (1993), 63–67.

57. G. Faglia, E. Comini, A. Cristalli, G. Sberveglieri, and L. Dori. "Very low power consumption micromachined CO sensors", *Sensors and Actuators* **B55** (1999), 140–146.

58. W.Y. Chung, J.W. Lim, D.D. Lee, N. Miura, and N. Yamazoe. "Thermal and gas-sensing properties of planar-type micro gas sensor", *Sensors and Actuators* **B64** (2000), 118–123.

59. R.E. Cavicchi, R.M. Walton, M. Aquino-Class, J.D. Allen, and B. Panchapakesan. "Spin-on nanoparticle tin oxide for microhotplate gas sensors", *Sensors and Actuators* **B77** (2001), 145–154.

60. I. Jimenez, A. Cirera, A. Cornet, J.R. Morante, I. Grácia, and C. Cané. "Pulverisation method for active layer coating on microsystems", *Sensors and Actuators* **B84** (2002), 78–82.

61. A. Heilig, N. Bârsan, U. Weimar, M. Schweizer-Berberich, J.W. Gardner, and W. Göpel. "Gas identification by modulating temperatures of SnO₂-based thick film sensors", *Sensors and Actuators* **B43** (1997), 45–51.

62. D. Vincenzi, M.A. Butturi, V. Guidi, M.C. Carotta, G. Martinelli, V. Guarnieri, S. Brida, B. Margesin, F. Giacomozzi, M. Zen, G.U. Pignatel, A.A. Vasiliev, and A.V. Pisliakov. "Development of a low-power thick-film gas sensor deposited by screen-printing technique onto a micromachined hotplate", *Sensors and Actuators* **B77** (2001), 95–99.

63. B. Riviere, J.-P. Viricelle, and C. Pijolat. "Development of tin oxide material by screen-printing technology for micro-machined gas sensors", *Sensors and Actuators* **B93** (2003), 531–537.

64. M. Heule, S. Vuillemin, and L.J. Gauckler. "Powder-based ceramic meso- and microscale fabrication processes", *Advanced Materials* **15** (2003), 1237–1245.

65. N. Bârsan and U. Weimar. "Conduction model of metal oxide gas sensors", *Journal of Electroceramics* **7** (2001), 143–167.

66. N. Bârsan and U. Weimar. "Understanding the fundamental principles of metal oxide based gas sensors; the example of CO sensing with SnO₂ sensors in the presence of humidity", *Journal of Physics: Condensed Matter* **15** (2003), R813-R839.

67. J. Cerdà, A. Cirera, A. Vilà, A. Cornet, and J.R. Morante. "Deposition on micromachined silicon substrates of gas sensitive layers obtained by a wet chemical route: a CO/CH₄ high performance sensor", *Thin Solid Films* **391** (2001), 265–269.

68. J. Kappler, N. Bârsan, U. Weimar, A. Diéguez, J.L. Alay, A. Romano-Rodriguez, J.R. Morante, and W. Göpel. "Correlation between XPS, Raman and TEM measurements and the gas sensitivity of Pt and Pd doped SnO₂ based gas sensors", *Fresenius' Journal of Analytical Chemistry* **361** (1998), 110–114.

69. U. Weimar and W. Göpel. "AC measurements on tin oxide sensors to improve selectivities and sensitivities", *Sensors and Actuators* **B26–27** (1995), 13–18.

70. R.E. Cavicchi, J.S. Suehle, K.G. Kreider, M. Gaitan, and P. Chaparala. "Optimized temperature-pulse sequences for the enhancement of chemically specific response patterns from micro-hotplate gas sensors", *Sensors and Actuators* **B33** (1996), 142–146.

71. M. Jaegle, J. Wöllenstein, T. Meisinger, H. Böttner, G. Müller, T. Becker, and C. Bosch-von Braunmühl. "Micromachined thin film SnO$_2$ gas sensors in temperature-pulsed operation mode", *Sensors and Actuators* **B57** (1999), 130–134.

72. M. Schweizer-Berberich, M. Zdralek, U. Weimar, W. Göpel, T. Viard, D. Martinez, A. Seube, and A. Peyre-Lavigne. "Pulsed mode of operation and artificial neural network evaluation for improving the CO selectivity of SnO$_2$ gas sensors", *Sensors and Actuators* **B65** (2000), 91–93.

73. T.A. Kunt, T.J. McAvoy, R.E. Cavicchi, and S. Semancik. "Optimization of temperature programmed sensing for gas identification using micro-hotplate sensors", *Sensors and Actuators* **B53** (1998), 24–43.

74. T. Takada. "A new method for gas identification using a single semiconductor sensor", *Sensors and Actuators* **B52** (1998), 45–52.

75. A. Heilig, N. Bârsan, U. Weimar, and W. Göpel. "Selectivity enhancement of SnO$_2$ gas sensors: simultaneous monitoring of resistances and temperatures", *Sensors and Actuators* **B58** (1999), 302–309.

76. T. Takada. "Temperature drop of semiconductor gas sensor when exposed to reducing gases – simultaneous measurement of changes in sensor temperature and in resistance", *Sensors and Actuators* **B66** (2000), 1–3.

77. D. Barrettino, M. Graf, M. Zimmermann, A. Hierlemann, and H. Baltes. "A Smart Single-chip Microhotplate-based Chemical Sensor System in CMOS Technology", *Proc. IEEE International Symposium on Circuits and Systems (ISCAS)*, Phoenix, AZ, USA (2002) Vol. 2, 157–160.

78. M.Y. Afridi, J.S. Suehle, M.E. Zaghloul, D.W. Berning, A.R. Hefner, S. Semancik, and R.E. Cavicchi. "A Monolithic Implementation of Interface Circuitry for CMOS Compatible Gas-Sensor System", *Proc. IEEE International Symposium on Circuits and Systems (ISCAS)*, Phoenix, AZ, USA (2002) Vol. 2, 732–735.

79. M.Y. Afridi, J.S. Suehle, M.E. Zaghloul, D.W. Berning, A.R. Hefner, R.E. Cavicchi, S. Semancik, C.B. Montgomery, and C.J. Taylor. "A monolithic CMOS microhotplate-based gas sensor system", *IEEE Sensors Journal* **2** (2002), 644–655.

80. M. Graf, D. Barrettino, M. Zimmermann, A. Hierlemann, and H. Baltes. "CMOS Single-chip Metal Oxide Gas Sensing System Based on Microhotplates", *Proc. 9th International Meeting on Chemical Sensors (IMCS)*, Boston, MA, USA, (2002), 141.

81. M. Graf, D. Barrettino, M. Zimmermann, C. Hagleitner, A. Hierlemann, and H. Baltes, S. Hahn, N. Bârsan, and U. Weimar. "CMOS Monolithic Metal-Oxide Sensor System Comprising a Microhotplate and Associated Circuitry", *IEEE Sensors Journal* **4** (2004), 9–16.

82. J. Kappler. *Characterisation of high-performance SnO$_2$ gas-sensors for CO detection by in situ techniques*, Ph.D. thesis, University of Tübingen, Shaker-Verlag, Germany (2001).

83. S. Hahn. *SnO$_2$ thick film sensors at ultimate limits: Performance at low O$_2$ and H$_2$O concentrations; Size reduction by CMOS technology*, Ph.D. thesis, University of Tübingen, Germany (2002).

84. S. Harbeck, A. Szatvanyi, N. Bârsan, U. Weimar, and V. Hoffmann. "DRIFT studies of thick film un-doped and Pd-doped SnO$_2$ sensors: temperature changes effect and CO detection mechanism in the presence of water vapour", *Thin Solid Films* **436** (2003), 76–83.

85. V.A. Henrich and P.A. Cox. *The Surface Science of Metal Oxides*, University Press, Cambridge, UK (1994), 312–316.

86. J. Kappler, N. Bârsan, U. Weimar, and W. Göpel. "Influence of water vapour on nanocrystalline SnO$_2$ to monitor CO and CH$_4$", *Proc. Eurosensors XI*, Warschau, Poland (1997), 1177–1180.

87. N. Bârsan, A. Heilig, J. Kappler, U. Weimar, and W. Göpel. "CO-water interaction with Pd-doped SnO$_2$ gas sensors: simultaneous monitoring of resistances and work functions", *Proc. Eurosensors XIII*, The Hague, Netherlands (1999), 183–184.

88. S.H. Hahn, N. Bârsan, and U. Weimar. "Investigation of CO/CH$_4$ mixture measured with differently doped SnO$_2$ sensors", *Sensors and Actuators* **B78** (2001), 64–68.

89. U. Weimar. *Gas Sensing with Tin Oxide: Elementary Steps and Signal Transduction*, Habilitation thesis, University of Tübingen, Germany (2001).

90. R. Ionescu, A. Vancu, C. Moise, and A. Tomescu. "Role of water vapour in the interaction of SnO$_2$ Gas Sensors with CO and CH$_4$", *Sensors and Actuators* **B61** (1999), 39–42.

91. D. Barrettino, M. Graf, M. Zimmermann, C. Hagleitner, A. Hierlemann, and H. Baltes. "A Smart Single-Chip Micro-Hotplate-Based Gas Sensor System in CMOS-Technology", *Analog Integrated Circuits and Signal Processing* **39** (2004), 275–287.

92. N.R. Swart and A. Nathan. "Design Optimization of Integrated Microhotplates", *Sensors and Actuators* **A43** (1994), 3–10.

93. S. Möller, J. Lin, and E. Obermeier. "Material and Design Considerations for Low-Power Microheater Modules for Gas-Sensor Applications", *Sensors and Actuators* **B25** (1995), 343–346.

94. C. Rossi, E. Scheid, and D. Estève. "Theoretical and experimental study of silicon micromachined microheater with dielectric stacked membranes", *Sensors and Actuators* **A63** (1997), 183–189.

95. M. Dumitrescu, C. Cobianu, D. Lungu, D. Dascalu, A. Pascu, S. Kolev, and A. van den Berg. "Thermal simulation of surface micromachined polysilicon hot plates of low power consumption", *Sensors and Actuators* **A76** (1999), 51–56.

96. D. Briand, S. Heimgartner, M.A. Grétillat, B. van der Schoot, and N.F. de Rooij. "Thermal optimization of micro-hotplates that have a silicon island.", *Journal of Micromechanics and Microengineering* **12** (2002), 971–978.

97. P. Ruther, M. Ehmann, T. Lindemann, and O. Paul. "Dependence of the Temperature Distribution in Micro Hotplates on Heater Geometry and Heating Mode", *Proc. IEEE Transducers '03*, Boston, MA, USA (2003), 73–76.

98. A. Pike and J.W. Gardner. "Thermal modelling and characterisation of micropower chemoresistive silicon sensors", *Sensors and Actuators* **B45** (1997), 19–26.

99. F. Solzbacher, T. Doll, and E. Obermeier. "A comprehensive analytical and numerical analysis of transient and static micro hotplate characteristics", *Proc. IEEE Transducers'03*, Boston, MA, USA (2003), 1856–1859.

100. J. Hildenbrand, J. Wöllenstein, E. Spiller, G. Kuhner, H. Böttner, G. Urban, and J.G. Korvink. "Design and fabrication of a novel low cost hotplate micro gas sensor", *Proceedings of the SPIE (The International Society for Optical Engineering)* **4755** (2002), 191–199.

101. T. Bechthold, E.B. Rudnyi, J.G. Korvink, M. Graf, and A. Hierlemann. "Connecting heat transfer macromodels for MEMS-array structures". *Journal of Micromechanics and Microengineering* **15** (2005), 1205–1214.

102. T. Bechthold, E.B. Rudnyi, and J.G. Korvink. "Dynamic electro-thermal simulation of microsystems – a review", *Journal of Micromechanics and Microengineering* **15** (2005), R17–R31.

103. M. von Arx, O. Paul, and H. Baltes. "Process-dependent thin-film thermal conductivities of thermal CMOS MEMS", *Journal of Microelectromechanical Systems* **9**, (2000), 136–145.

104. S. Hafizovic and O. Paul. "Temperature-dependent thermal conductivities of CMOS layers by micromachined thermal van der Pauw test structures", *Sensors and Actuators* **A97–98** (2002), 246–252.

105. G.A. Slack. "Thermal Conductivity of Pure + Impure Silicon, Silicon Carbide + Diamond", *Journal of Applied Physics* **35** (1964), 3460–3466.

106. C.J. Glassbrenner and G.A. Slack. "Thermal Conductivity of Silicon and Germanium from 3 °K to the Melting Point", *Physical Review* **A 134** (1963), 1058–1069.
107. F.P. Incropera and D.P. De Witt. *Introduction to heat transfer*, John Wiley, New York, USA (2002).
108. D. Gibson, H. Carter, and C. Purdy. "The Use of Hardware Description Languages in the Development of Microelectromechanical Systems", *Analog Integrated Circuits and Signal Processing* **28** (2001), 173–180.
109. B. Kloeck, S.D. Collins, N.F. de Rooij, and R.L. Smith. "Study of electrochemical etch-stop for high-precision thickness control of silicon membranes", *IEEE Transactions on Electron Devices* **36** (1989), 663–669.
110. T. Müller, M. Brandl, O. Brand, and H. Baltes. "An industrial CMOS process family adapted for the fabrication of smart silicon sensors", *Sensors and Actuators* **A84** (2000), 126–133.
111. N. Kerness. *CMOS-Based Calorimetric Chemical Microsensors*, Ph.D. thesis No. 14839, ETH Zurich, Switzerland, Hartung-Gorre-Verlag, Konstanz, Germany (2002).
112. M. von Arx. *Thermal Properties of CMOS Thin Films*, Ph.D. thesis No. 12743 ETH Zurich, Switzerland, Hartung-Gorre-Verlag, Konstanz, Germany (1998).
113. T.H. Ning (Editor). *Properties of Silicon*, INSPEC, London and New York, UK and USA (1988).
114. Gmelin and Zinn. "Gmelin Handbuch der anorganischen Chemie", 8th Edition, Vol. CI, (1972).
115. R. Jurischka. *Microhotplate-basierte Metalloxid-Gassensoren in CMOS-Technologie*, Diploma thesis, Institute for Microsystem Technology, University of Freiburg, Germany (2002).
116. M. Graf, R. Jurischka, D. Barrettino, and A. Hierlemann. "3D nonlinear modeling of microhotplates in CMOS technology for use as metal-oxide-based gas sensors", *Journal of Micromechanics and Microengineering* **15** (2005), 190–200.
117. V. Lysenko, S. Perichon, B. Remaki, and D. Barbier. "Thermal isolation in microsystems with porous silicon", *Sensors and Actuators* **A99** (2002), 13–24.
118. M. Ehmann, P. Ruther, M. von Arx, and O. Paul. "Operation and short term drift of polysilicon-heated CMOS microstructures at temperatures up to 1200 K", *Journal of Micromechanics and Microengineering* **11** (2001), 397–401.
119. L.Y. Sheng, C. de Tandt, W. Ranson, and R. Vounckx. "Reliability aspects of thermal micro-structures implemented on industrial 0.8 μm CMOS chips", *Microelectronics Reliability* **41** (2001), 307–315.
120. J. Puigcorbé, A. Vilà, J. Cerdà, A. Cirera, I. Grácia, C. Cané, and J.R. Morante. "Thermomechanical analysis of micro-drop coated gas sensors", *Sensors and Actuators* **A97–98** (2002), 379–385.
121. A.K. Stamper and S.L. Pennington. "Characterization of Plasma-Enhanced Chemical Vapor Deposited Nitride Films Used in Very Large Scale Integrated Applications", *Journal of The Electrochemical Society*, **140** (1993), 1748–1752.
122. U. Münch, D. Jaeggi, N. Schneeberger, A. Schaufelbühl, O. Paul, H. Baltes, and J. Jasper. "Industrial fabrication technology for CMOS infrared sensor arrays", *Dig. Tech. Papers Transducers '97*, Vol. 1, Chicago, IL, USA, (1997), 205–208.
123. J. Goetz. "Sensors that can take the heat – Part 1.", *Sensors* **17** (2000), Issue 6, 20–38.
124. J. Goetz. "Sensors that can take the heat – Part 2.", *Sensors* **17** (2000), Issue 7, 52–96.
125. M. Graf, S.K. Müller, D. Barrettino, and A. Hierlemann. "Transistor heater for microhotplate-based metal-oxide microsensors", *IEEE Electron Device Letters* **26** (2005), 295–297.

126. S. Müller. *CMOS-Micro-Hotplate with MOS Transistor Heater for Integrated Metal Oxide Gas Sensors*, Diploma thesis, ETH Zurich, Switzerland (2002).

127. H. Shichman and D. Hodges. "Modelling and simulation of insulated-gate field effect transistor switching circuits", *IEEE Journal of Solid-State Circuits* **SC3** (1968), 285–289.

128. D. Foty. *MOSFET Modeling with Spice: Principles and Practice*, Prentice Hall PTR, New Jersey, USA (1997).

129. W. Liu, X. Jin, J. Chen, M.-C. Jeng, Z. Liu, Y. Cheng, K. Chen, M. Chan, K. Hui, J. Huang, R. Tu, P.K. Ko, and C. Hu. "BSIM3.v3.2.2 MOSFET Model – User's Manual", *http://www-device.EECS.Berkeley.EDU/~bsim3/intro.html*, University of California, Berkeley, CA, USA (1999).

130. D. Barrettino, M. Graf, S. Taschini, M. Zimmermann, C. Hagleitner, A. Hierlemann, and H. Baltes. "Hotplate-based conductometric monolithic CMOS gas sensor system", *Proc. IEEE VLSI Symposium*, Kyoto, Japan, (2003), 303–306.

131. M. Graf, D. Barrettino, S. Taschini, C. Hagleitner, A. Hierlemann, and H. Baltes. "Metal-Oxide-Based Monolithic Complementary Metal Oxide Semiconductor Gas Sensor Microsystem", *Analytical Chemistry* **76** (2004), 4437–4445.

132. D. Barrettino, M. Graf, S. Taschini, S. Hafizovic, C. Hagleitner, and A. Hierlemann. "CMOS monolithic metal-oxide gas sensor Microsystems", *IEEE Sensors Journal* **6** (2006), 276–286.

133. J.J.F. Rijns, "CMOS Low-Distortion High-Frequency Variable-Gain Amplifier", *IEEE Journal of Solid-State Circuits* **31** (1996), 1029–1034

134. R.J. van de Plassche, "A Wide-Band Monolithic Instrumentation Amplifier", *IEEE Journal of Solid-State Circuits* **10** (1975), 424–431.

135. S.D. Willingham, K.W. Martin, and A. Ganesan, "A BiCMOS Low-Distortion 8-MHz Low-Pass Filter", *IEEE Journal of Solid-State Circuits*, Vol. **28** (1993), 1234–1245.

136. R.J. Baker, H.W. Li and D.E. Boyce, *CMOS Circuit Design, Layout, and Simulation*, IEEE Press, USA, (1998).

137. R.J. Wiegerink, *Analysis and Synthesis of MOS Translinear Circuits*, Kluwer, 1993.

138. J.J.F. Rijns, "CMOS Low-Distortion High-Frequency Variable-Gain Amplifier", *IEEE Journal of Solid-State Circuits* **31** (1996), 1029–1034.

139. C. Hung, M. Ismail, K. Halonen, and V. Porra, "Low-Voltage Rail-to-Rail CMOS Differential Difference Amplifier", *Proc. IEEE International Symposium on Circuits and Systems*, Vol. 1 (1997), 145–148.

140. H. Baltes, O. Brand, and M. Waelti. "Packaging of CMOS MEMS", *Microelectronics Reliability* **40** (2000), 1255–1262.

141. M. Graf, D. Barrettino, K.-U. Kirstein, and A. Hierlemann. "CMOS microhotplate sensor system for operating temperatures up to 500 °C", *Sensors and Actuators* **B117** (2006), 346–352.

142. A. Hierlemann, M. Schweizer-Berberich, U. Weimar, G. Kraus, A. Pfau, and W. Göpel. "Pattern Recognition and Multicomponent Analysis", *Sensors Update, Vol. 2*, VCH, Weinheim, Germany (1996), 121–180.

143. R. Gutierrez-Osuna. "Pattern analysis for machine olfaction: a review", *IEEE Sensors Journal* **2** (2002), 189–202.

144. M. Graf, D. Barrettino, P. Käser, J. Cerdà, A. Hierlemann, and H. Baltes, "Smart single-chip CMOS microhotplate array for metal-oxide-based gas sensors" *Proc. IEEE Transducers '03*, Boston, MA, USA (2003), 123–126.

145. D. Barrettino, M. Graf, W.H. Song, K.-U. Kirstein, A. Hierlemann, and H. Baltes. "Hotplate-Based Monolithic CMOS Microsystems for Gas Detection and Material Characterization for Operating Temperatures up to 500 °C", *IEEE Journal of Solid-State Circuits* **39** (2004), 1202–1207.

146. D. Barrettino. Graf, S. Taschini, S. Hafizovic, C. Hagleitner, and A. Hierlemann. "A single-chip CMOS micro-hotplate array for hazardous-gas detection and material characterization", *Proc. IEEE International Solid-State Circuits Conference (ISSCC)*, San Francisco, CA, USA (2004), 311–312.

147. U. Frey, M. Graf, S. Taschini, K.-U. Kirstein, C. Hagleitner, A. Hierlemann, and H. Baltes. "A Digital CMOS Micro-Hotplate Array for Analysis of Environmentally Relevant Gases", *Proc. IEEE European Solid-State Circuits Conference (ESSCIRC)*, Leuven, Belgium (2004), 299–302.

148. U. Frey, M. Graf, S. Taschini, K.-U. Kirstein, and A. Hierlemann. "A Digital CMOS Architecture for a Micro-Hotplate Array", *IEEE Journal of Solid-State Circuits* (2007), 42, 441–450.

149. M. Graf, U. Frey, S. Taschini, and A. Hierlemann. "Microhotplate-Based Sensor Array System for the Detection of Environmentally Relevant Gases", *Analytical Chemistry* **78** (2006), 6801–6808.

150. M. Graf, U. Frey, P. Reichel, S. Taschini, N. Barsan, U. Weimar, and A. Hierlemann. "Monitoring of environmentally monolithic metal-oxide relevant gases by a digital microsensor array." Proc. IEEE Sensors Conference, Vienna, Austria (2004), 776–779.

Subject Index

Printing: Krips bv, Meppel
Binding: Stürtz, Würzburg